M63

Aquifer Storage and Recovery

American Water Works Association

Manual of Water Supply Practices—M63

Aquifer Storage and Recovery

Sr. Manager, Editorial, Development and Production: Gay Porter DeNileon
Senior Manuals Specialist: Molly Beach
Production: TIPS Technical Publishing, Inc.

If you find errors in this manual, please email books@awwa.org. Possible errata will be posted at www.awwa.org/resources-tools/resource.development.groups/manuals-program.aspx.

Library of Congress Cataloging-in-Publication Data

Bloetscher, Frederick, author.
Aquifer storage and recovery / Frederick Bloetscher. — First edition.
pages cm. — (Manual of water supply practices ; M63)
Includes bibliographical references and index.
ISBN 978-1-62576-104-0
1. Aquifer storage recovery-—United States. I. Title. II. Series: AWWA manual ; M63.
TD404.5.B55 2015
628.1'3—dc23
2015011138

ISBN-13 978-1-62576-104-0
eISBN-13 978-1-61300-340-4

Printed in the United States of America
American Water Works Association
6666 West Quincy Avenue
Denver, CO 80235-3098
awwa.org

Contents

Figures

This page intentionally blank.

Tables

This page intentionally blank.

Foreword

M63, *Aquifer Storage and Recovery,* is a new addition to the Manuals of Water Supply Practices produced by the American Water Works Association. This manual is the culmination of two years of effort by a subset of the members of the Groundwater Resources Committee. M63 was written to provide the reader with a general understanding of the principles involved with aquifer storage and recovery (ASR). The contents of this manual focus on the concept, the general rules as they are applied nationally and by state, basic design criteria, and the results of an inventory of ASR well sites nationally. Most of these wells are used to store potable water, and as a result are intended to require only limited treatment prior to use in the water distribution system. At the same time, a quarter of the wells are currently inactive for a variety of reasons.

The intention of the ASR subcommittee was to create a document that provides a general overview, without the detailed mathematical analyses and equations that are available in many other groundwater textbooks and references, including M21, *Groundwater,* AWWA Standard A100 Water Wells, and the AWWA publication *Siting, Drilling, and Construction of Water Wells* (Bloetscher et al. 2007). This manual will provide operators and engineering staff with an understanding of ASR that will help them make decisions on investigations and installations when problems or the need to expand supplies arise. ASR is a new frontier for many states and utilities. We do know that aquifers and wells vary across the country, ASR does not work everywhere, and it is not a panacea for all water-supply issues. But ASR can be a useful tool in the utility manager's toolbox.

The scope of this manual is as follows:

Chapter 1 is an overview of aquifer injection programs and how they are differentiated from ASR systems.

Chapter 2 is an overview of the rules set forth by the Safe Drinking Water Act (SDWA) and the subsequent delegation of the enforcement authority to states. Because a number of states do not have underground injection control (UIC) primary enforcement authority, USEPA directly implements the UIC programs (including ASR wells) for these states. ASR rules change periodically so only an overview has been provided.

Chapter 3 summarizes the experience in the United States with ASR, including locations and characteristics, with a much-improved expansion from previous ASR surveys.

Chapter 4 outlines the challenges of ASR projects, including metals leaching, plugging, and recovery issues.

Chapter 5 outlines the basic guidelines for design and construction of ASR wells, with emphasis on the testing protocol and expectations that must be conveyed to the client. ASR testing takes time (up to five years), so for ASR to be successful, implementation must be carefully considered and planned. Chapter 5 also outlines the evaluation methods and parameters for ASR wells, including:

- Aquifer parameters
- Transmissivity and storage coefficient
- Collection of test data

- Analysis
- Well design

Well losses, yield, and modeling techniques are covered in detail in other publications, including those mentioned previously, and are not covered in detail herein.

Chapter 6 outlines issues associated with maintenance of ASR wells. Note, maintenance experience is limited because most wells are less than 20 years old.

Chapter 7 details specific examples of ASR experiences, both successful and those facing challenges since overcoming the challenge is often more informative than the successes.

Chapter 8 summarizes the manual and outlines how aquifer recharge can be accomplished by other means. Aquifer recharge is expected to be a topic for a future manual.

This manual should help operators and engineers gain enough background on the subject of ASR to improve their decision making in the United States. M63 should help these professionals answer many questions about complex aquifer systems and improve their response to problems. The Groundwater Resources Committee is hopeful that this manual will meet the industry needs of the new millennium.

Acknowledgments

The AWWA Technical & Educational Council, the Water Resource Sustainability Division, and the Groundwater Resources Committee gratefully acknowledge the contributions made by those volunteers who drafted, edited, and provided the significant and critical commentary essential to developing M63. The Technical Review Board members dedicated many hours in the final review of this manual to ensure its overall technical quality, consistency, and accuracy.

Technical Review Board Members

Frank Getchell, Leggette Brashears & Graham Inc., Upper Saddle River, N.J.
Lindsey Geiger, American Water Works Association, Denver, Colo.
Stephen A. Hubbs, University of Louisville, Louisville, Ky.
Albert Muniz, Hazen and Sawyer P.C., Boca Raton, Fla.
Joseph M. Nattress, CH2M, Philadelphia, Pa.
Susan Roberts, Vice Chair, Abengoa-Abeinsa, Chesterfield, Mo.

Contributors

Frederick Bloetscher, Chair, Florida Atlantic University, Hollywood, Fla.
Fred M. Blumberg, ARCADIS U.S. Inc., San Antonio, Texas
Richard E. Cowles, Malcolm Pirnie, Sarasota, Fla.
Frank Getchell, Leggette Brashears & Graham Inc., Upper Saddle River, N.J.
Jeffrey W. Henson, Black & Veatch, Kansas City, Mo.
Stephen A. Hubbs, University of Louisville, Louisville, Ky.
Timothy P. Karlstrand, Carollo Engineers, Walnut Creek, Calif.
Mark C. Lucas, CH2M, Fort Lauderdale, Fla.
Bryan B. McDonald, CH2M, Gainesville, Fla.
Albert Muniz, Hazen and Sawyer P.C., Boca Raton, Fla.
Joseph M. Nattress, CH2M, Philadelphia, Pa.
Peter E. Petersen, Milwaukee Water Works, Milwaukee, Wis.
Susan Roberts, Vice Chair, Abengoa-Abeinsa, Chesterfield, Mo.
Laura L. Simons, New Jersey American Water, Voorhees, N.J.
Kevin Spencer, R. W. Harden & Associates Inc., Austin, Texas
Timothy M. Thomure, HDR Engineering, Phoenix, Ariz.

This manual required the support of many individuals.
These people provided data, information, or contacts:

Bassad Ahmadi, Valley Water, Fresno, Calif.
Randy Albright, R.G., L.H.G., Portland Water Bureau, Portland, Ore.
Brian P. Austin, Wisconsin Department of Natural Resources, Madison, Wis.
Steven G. Bayley, City of Tracy, Tracy, Calif.
Marcel Belaval, USEPA Region I, Boston, Mass.
Craig Boomgaard, USEPA Region 7, Denver, Colo.

Bruce Bunch, City of Fresno, Fresno, Calif.
Gus Cicala, East Bay MUD Library Services, Oakland, Calif.
David Cole, R.G., Oregon DEQ Northwest Region, Portland, Ore.
Brian A. Dickinson, Elsinore Valley Municipal Water District, Lake Elsinore, Calif.
Alan R. Dulaney, City of Peoria Utilities Department, Peoria, Ariz.
Kelley A. Dyer, City of Santa Barbara, Santa Barbara, Calif.
Denice Enochs, Elsinore Valley Municipal Water District, Lake Elsinore, Calif.
Robert J. Eustice, Florida Atlantic University, West Palm Beach, Fla.
William Ferguson, City of Santa Barbara, Santa Barbara, Calif.
Anna L. Garcia, Mojave Water Agency, Apple Valley, Calif.
Robert C. Garcia, City of Peoria, Peoria, Ariz.
Michael E. Garrod, Sweetwater Authority, Chula Vista, Calif.
Carolyn German, Irvine Ranch Water District, Irvine, Calif.
Gary M. Gin, City of Phoenix, Water Services Department, Phoenix, Ariz.
Mary Shaleen Hansen, Washington Department of Ecology, Water Resources Program, Olympia, Wash.
Gary Hilliard, Los Angeles County Department of Public Works, Stanton, Calif.
John Izbicki, USGS, San Diego, Calif.
Elizabeth Janes, USEPA Region 9, San Francisco, Calif.
Matt Katen, Zone 7 Water Agency, Livermore, Calif.
Dave J. Kaunisto, E. Cherry Creek Valley Water and Sanitation, Aurora, Colo.
Jordan Kear, Kear Groundwater, Santa Barbara, Calif.
Karl Kennedy, Calvin Giordano & Associates, Fort Lauderdale, Fla.
J. Bart King, Los Angeles County Department of Public Works, Huntington Beach, Calif.
Susan King, Sweetwater Authority, Chula Vista, Calif.
Nicole Lance (Dailey), City of Gilbert, Gilbert, Ariz.
Jeffrie Madland, City of Camarillo, Camarillo, Calif.
Craig Maples, Chesapeake Public Utilities, Chesapeake, Va.
Robert Marks, Pueblo Water Resources, Ventura, Calif.
Joseph May, FDEP, Southeast District, West Palm Beach, Fla.
Andy McDonald, N.J. Department of Environmental Protection, Trenton, N.J.
Gary McSmith, Cape Fear Public Utility Authority, Wilmington, N.C.
Kirk D. Medina, North Las Vegas Utility, Las Vegas, Nev.
Michelle Moreno, Arizona Dept. of Water Resources, Phoenix, Ariz.
Tom M. Morris, ASR Systems, Las Vegas, Nev.
Behrooz Mortazavi, Eastern Municipal Water District, Perris, Calif.
Jim Murphy, Chief, N.J. Department of Environmental Protection, Trenton, N.J.
Dave Nazy, Washington Department of Ecology, Water Resources Program, Olympia, Wash.
Richard M. Noble, P.E., Hansen, Allen & Luce Inc., American Fork, Utah
Joe Oliver, Monterey Peninsula Water Management District, Monterey, Calif.
Alan E. Packard, Jordan Valley Water Conservancy District, West Jordan, Utah
Tony Perez, City of Fresno, Fresno, Calif.
Terry Pilawski, N.J. Department of Environmental Protection, Trenton, N.J.

David Pyne, ASR Systems, Gainesville, Fla.

Cher Quinones, Development & Records Coordinator, EVMWD, Lake Elsinore, Calif.

Laurie Richards, Carpinteria Valley Water District, Carpinteria, Calif.

Claudia Sandoval, Irvine Ranch Water District, Irvine, Calif.

Chi Ho Sham, The Cadmus Group Inc., Waltham, Mass.

Scotty D. Sorrells, Tennessee Dept. of Environment and Conservation, Groundwater Management Section, Nashville, Tenn.

Kimberly Swan, Clackamas River Water, Clackamas, Ore.

Linda Townes, Jordan Valley Water Conservancy District, West Jordan, Utah

Dana Walker, City of Phoenix, Phoenix, Ariz.

Daniel Wendell, Groundwater Dynamics, Oakland, Calif.

Kathy Wright, State of Delaware, DNREC, Wilmington, Del.

Mark Wuttig, CH2M, Thousand Oaks, Calif.

Their help was invaluable in completing a full compilation of ASR projects. Our apologies to those missed for our failure to record their email or voice information.

Frederick Bloetscher, Ph.D., P.E., Chair,
Groundwater Resources Committee

This page intentionally blank.

Acronyms

AFA	acre-ft annually
AGWT/HC/FGS	American Groundwater Trust/Hydrogeology Consortium/Florida Geological Survey
ASCE-EWRI	American Society of Civil Engineers – Environmental & Water Resources Institute
ASR	aquifer storage and recovery
CAP	Central Arizona Project
$CBOD_5$	5-day carbonaceous biological oxygen demand
CERP	Comprehensive Everglades Restoration Plan
CFR	Code of Federal Regulations
DBPs	disinfection by-products
DWR	Department of Water Resources
FAC	Florida Administrative Code
fpd	feet per day
GEFA	Georgia Environmental Finance Authority
gpd/ft	gallons per day/foot
gpd/ft^2	gallons per day/square foot
gpm	gallons per minute
gpm/ft	gallons per minute/foot
GSWSA	Grand Strand Water and Sewer Authority
H^+	hydrogen ion
HAAs	haloacetic acids
ID	internal diameter
K	hydraulic conductivity
LVV	Las Vegas Valley
MCL	maximum contaminant levels
mgd	million gallons per day
mi^2	square miles
MLGW	Memphis Light Gas and Water
Mn	manganese
NOM	natural organic matter
NPSH	net positive suction head
NTU	nephelometric turbidity units
OD	outside diameter
OH^-	hydroxide ion
ORP	oxidation–reduction potential
PAD	prior appropriation doctrine
PRRWSF	Peace River Regional Water Supply Facility

PVC	polyvinyl chloride
POS	plan of study
SCADA	supervisory control and data acquisition
SDWA	Safe Drinking Water Act
SFWMD	South Florida Water Management District
SWPP	source water protection program
TDH	total dynamic head
TDS	total dissolved solids
THMs	trihalomethanes
TOC	total organic carbon
TSS	total suspended solids
TSWB	Texas State Water Board
TTHM	total trihalomethanes
UIC	underground injection control
USDW	underground sources of drinking water
USEPA	United States Environmental Protection Agency
USGS	United States Geological Survey
VFD	variable-frequency drive
WCASF	Water Conservation Areas of South Florida
WEFTEC	Water Environment Federation Technical Exhibition and Conference
WWTP	wastewater treatment plant
ZOC	zone of contribution

Chapter 1

Groundwater Recharge and Storage Programs

In some areas of the world, vast amounts of treated or treatable water are being pumped below ground and stored beneath the earth's surface to preserve current water resources, prepare for future droughts, protect water resources, recharge wellfields, and store water for use at a later time to sustain development. Belowground storage holds an advantage over surface reservoirs because no evaporation losses occur. Water resources are also conserved because water not needed presently is stored, instead of being discharged into rivers, oceans, or other surface waterways.

Groundwater recharge and storage programs are divided into four categories: artificial aquifer creation, aquifer recharge, aquifer reclamation, and aquifer storage and recovery (ASR). All methods can be part of a water supplier's program of ensuring that sustainable water supplies are available for agricultural, environmental, and urban uses. Most methods are employed in arid or coastal areas or in areas with seasonal wet and dry periods, and also serve as a supplemental source in the event of emergency outages, low fire flows, and water quality issues (via blending). These strategies are employed in Australia, Europe, the Middle East, and the United States. At present, all of these strategies have proven cost effective and capable of successfully augmenting existing water sources for long-term water supply under certain conditions.

ARTIFICAL AQUIFER CREATION AND AQUIFER RECHARGE

Artificial aquifer creation technologies involve the introduction of large quantities of water into an aquifer or aquifer zone for retrieval down-gradient of the point of injection. Prior to injection, the aquifer is generally either devoid of water or contains low-quality water. In areas with limited demand, but where porosities or strata may allow for water storage, artificial aquifer recharge is applicable because it is specifically meant to encourage recharge and therefore requires high-quality water for injection. Because groundwater movement is generally slow, the recharge program may be able to supply either small quantities of water or supplement existing water during times of high demand that could cause aquifer stress.

Where an aquifer has been depleted by overpumping (mining of the aquifer), aquifer recharge, also termed artificial recharge, is a viable concept. This concept uses water of a given quality introduced at a point that enables existing water supplies to flow into a wellfield production zone. The aquifer head is raised enough to create a driving force that pushes water into the aquifer formations, where the water can be eventually pumped via a wellfield, which could be many miles away (Figures 1-1a and 1-1b).

Sometimes a surface area is simply flooded to create an artificial aquifer head at the point of recharge. The higher surface water head increases the percolation rate into the aquifer. One drawback with flooding areas is that surface water has a higher evapotranspiration rate than do other forms of aquifer recharge. However, downstream well fields benefit from the resulting higher water levels with increased total water supplies.

The Water Conservation Areas of South Florida (WCASF) are examples of aquifer recharge via flooding. The WCASF store water at higher-than-normal levels to force recharge of the Biscayne Aquifer system. The aquifer's natural gradient toward the coast allows the recharge to serve most of the South Florida utilities in some manner. Other such programs exist in Texas, Arizona, Utah, California, and other states, but artificial aquifer creation and aquifer recharge programs are outside the scope of this manual.

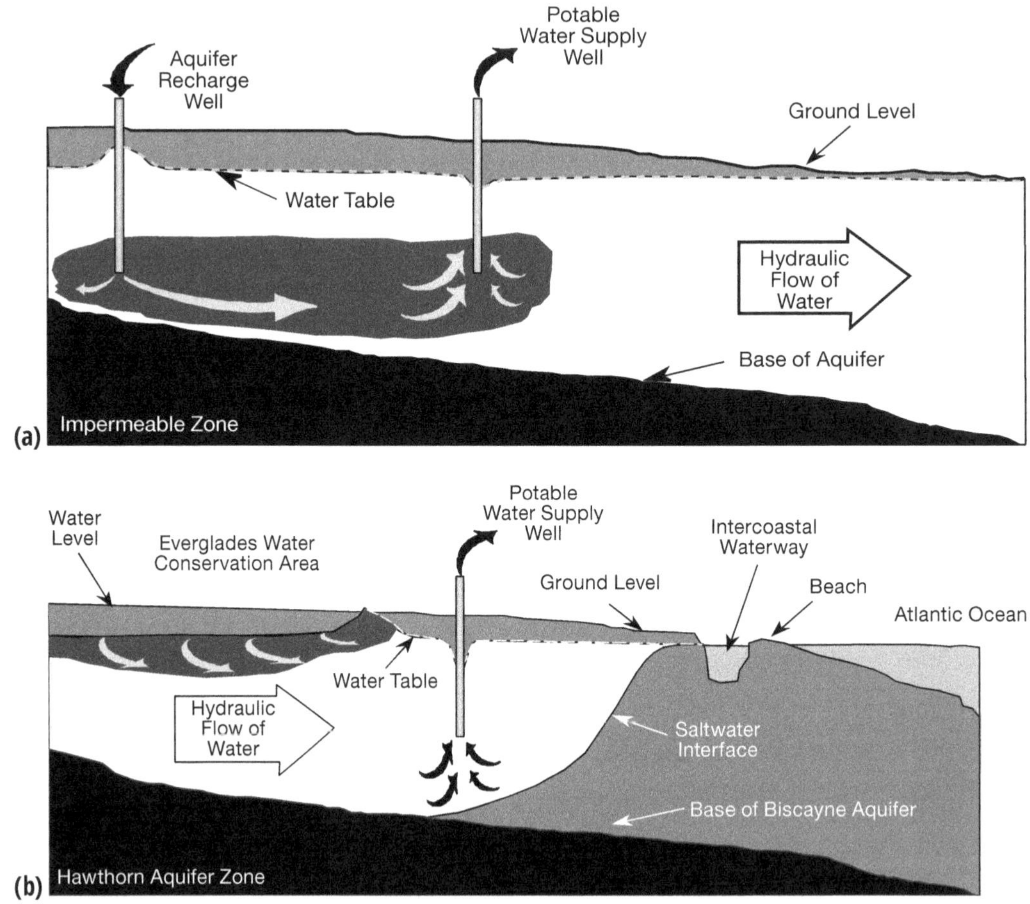

Figure 1-1 (a) Aquifer recharge; (b) Aquifer recharge via flooding

Source: *AWWA 2014*

AQUIFER RECLAMATION

Aquifer reclamation involves injecting large quantities of higher-quality waters into a compromised aquifer. One application of this technology is injection of freshwater into aquifer zones that have been compromised by brackish water (water containing 1,000 to 10,000 ppm of total dissolved solids) intruding into the formation (USEPA 2002). The freshwater serves to stabilize the water quality at a given chloride (isochlor) concentration or forces the brackish water to retreat toward the source—typically the ocean, but also lower formations where *upconing* is a problem. Upconing is the process by which saline water underlying freshwater in an aquifer rises upward into the freshwater zone as a result of pumping water from the freshwater zone. Aquifer reclamation has not been used in North America to a great extent except in California, but the technical concept has wide applicability in coastal areas with transmissive formations where saltwater intrusion has occurred or has the potential to occur, as a result of lowered surface water levels or where *connate water*—water trapped within the pores of rock formations—may be present. The method can also be applied to well fields where upconing of brackish or water containing high total dissolved solids from lower aquifer regions is a potential problem, thereby allowing withdrawals of larger quantities without inducing upconing.

Artificially-induced saline encroachment (intrusion) can be initiated when the aquifer head in a given area is sufficiently reduced by aquifer withdrawals, development over surface recharge areas reduces permeability (as a result of buildings and paving), or drainage canals discharge surface water that might otherwise percolate into the aquifer. Saltwater can continue to migrate inland over time, contaminating the aquifer. Initially the water movement is slow, but it accelerates as the saltwater draws closer to withdrawal (discharge) points. A hydrogeologically sound injection program, properly designed and constructed, can mitigate the landward migration of the saltwater. Under certain scenarios, the movement of the saltwater can be reversed (Figures 1-2a, 1-2b, 1-2c, and 1-2d). Chapter 11 of M21, *Groundwater* (AWWA 2014), and Bloetscher et al. (2005) discuss this concept in more detail.

AQUIFER STORAGE AND RECOVERY

That leaves us with the topic of this manual: aquifer storage and recovery, or ASR. ASR involves the management of water supplies for both potable (drinking water) and non-potable water supply systems. While the purpose of both aquifer reclamation and aquifer recharge is to sustain long-term demands in an existing or potential source, ASR is intended to store water until such time as it is needed to meet peak needs, long-term growth demands, or emergency conditions, or because of poor water quality.

More than 200 sites in 27 different states in the United States have either implemented or investigated ASR. Most existing systems involve storage of potable water, but a number of wells use untreated raw surface water or groundwater in an ASR system for later withdrawal and treatment. The goal of these projects is to store current water supplies for later use, both long and short term. As water supplies have become more limited in more areas of the country, additional interest has developed in the practice of incorporating ASR into potable water, irrigation, and reclaimed systems.

Employing ASR technology can increase the efficiency of water system operation. During a wet or low-demand time of the year, some or all of the unused water treatment plant capacity can be used to treat water and inject the treated water into an aquifer for future recovery and use. The potable water to be stored is injected into the aquifer (Figure 1-3) where it displaces the native groundwater in the aquifer. The storage zone should have good overlying confinement, underlying semi-confinement, and a transmissivity commensurate with the storage and recovery capacity goals of the program. In the ASR concept, native water is theoretically displaced and replaced in the injection zone,

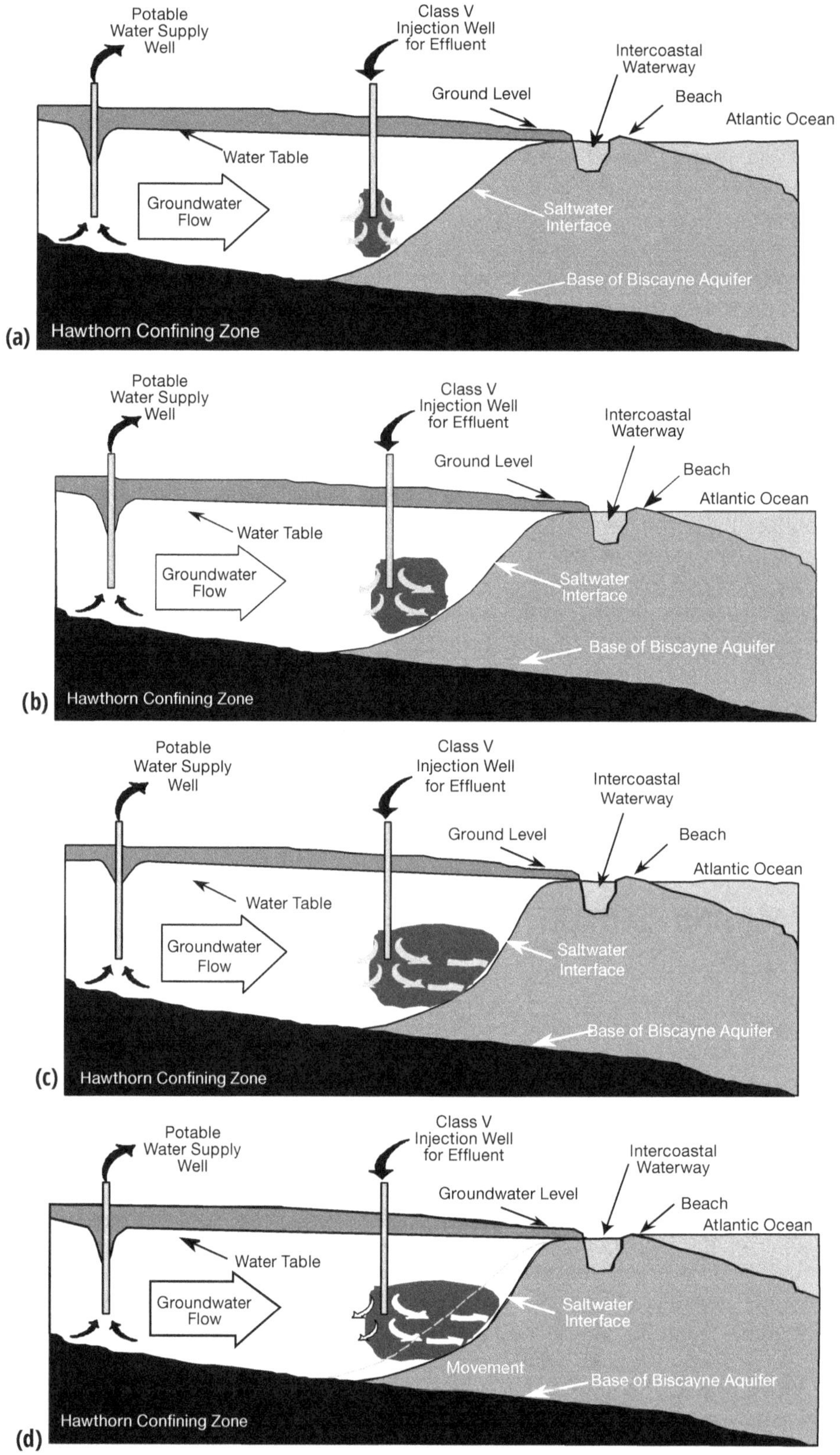

Figure 1-2 (a) Biscayne Aquifer reclamation after start of injection; (b) Biscayne Aquifer reclamation water movement after injection; (c) Biscayne Aquifer reclamation water movement toward saltwater after injection; (d) Biscayne Aquifer reclamation water movement pushing isochlor line toward the ocean

Source: *AWWA 2014*

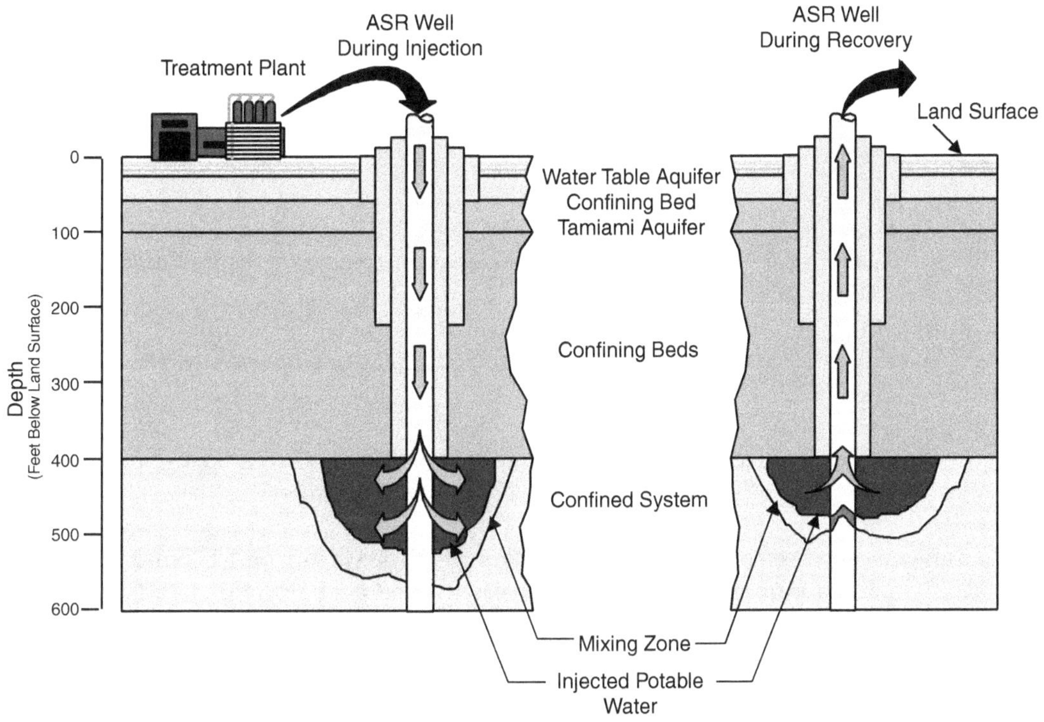

Figure 1-3 Aquifer storage and recovery conceptual diagram for brackish water aquifers where freshwater bubble creates mixing transition zone that increases with time

Source: *AWWA 2014*

as shown in the figures. With good underground conditions the injected water will hold together to create an underground bubble, depicted as a cone or half cylinder of fresh water in the native water. In theory, 100 percent recovery can be obtained from injection into a freshwater aquifer because of the minimal mixing zone caused by density stratification between the two waters, but geochemistry and other issues can frustrate this objective. Brackish aquifers are more challenging (see chapter 4).

The successful development of ASR projects depends upon both technological and regulatory factors. Both physical and chemical properties of the injected water may change after injection as a function of chemical reactions between water and the aquifer. The stored water has the potential to change as a result of

- increased aquifer pressure,
- temperature (increases or decreases),
- the potential for microorganism metabolization and secretions, and
- geochemical reactions.

Economic factors also impact the success of ASR projects. Technological and economic success depend on geology, excess water supply capacity (generally at least 1 mgd of excess capacity is needed, based on the data gathered from ASR sites across the country), and variations in water needs throughout the year. ASR is a water management strategy—it does not create water; it stores water underground instead. However, ASR can make managing supplies more effective under certain conditions. While ASR can be an attractive alternative to aboveground storage, the combined challenges of meeting requirements for drinking water storage and groundwater injection, and the need to establish ownership of the stored water, have created several challenges that must be addressed by regulatory agencies within each state/territory where ASR is used.

REFERENCES

American Water Works Association (AWWA). 2002. *Survey and Analysis of Aquifer Storage and Recovery (ASR) Systems and Associated Regulatory Programs in the United States*. Denver, CO: AWWA.

AWWA. 2014. Manual of Water Supply Practices M21, *Groundwater*, 4th ed. Denver, CO: AWWA.

Bloetscher, Frederick, Albert Muniz, and Gerhardt M. Witt. 2005. *Groundwater Injection: Modeling, Risks, and Regulations*. New York: McGraw-Hill.

Bloetscher, Frederick, Charles W. Walker, W. Kirk Martin, and Thomas M. Missimer. 1996a. Aquifer storage and recovery to meet peak water demands in Southwest Florida. *Proceedings of CONSERV96*, pp. 825–829. Denver, CO: AWWA.

Bloetscher, Frederick, Charles W. Walker, Patrick A. Davis, Albert Muniz, and Whitfield R. Van Cott. 1996b. Innovative use of reclaimed water to protect surficial drinking water supplies. *Florida Water Resource Conference Proceedings,* May 7, 1996.

Bloetscher, Frederick, and Gerhardt M. Witt. 1997. Aquifer storage and recovery as a means of resource management: A summary. *AWWA Annual Conference and Exposition Proceedings*, June 1997, Atlanta, Georgia. Denver, CO: AWWA.

USEPA (US Environmental Protection Agency). 2002. *A Lexicon of Cave and Karst Terminology with Special Reference to Environmental Karst Hydrology*. EPA/600/R-02/003. Washington, DC: USEPA.

Chapter 2

Regulatory Requirements

The United States Environmental Protection Agency (USEPA) is tasked with protecting underground sources of drinking water (USDWs) as a part of the Safe Drinking Water Act (SDWA), which includes the authority to regulate underground injection projects. A USDW is an aquifer containing groundwater with total dissolved solids (TDS) less than 10,000 mg/L (USEPA 2012). The Underground Injection Control (UIC) program—specific regulations promulgated at 40 CFR 144 to 148—protects USDWs from endangerment from injected fluid migration (USEPA 1983). Aquifers that are adjacent to the injection zone but are not USDWs are not subject to UIC regulations, but states may regulate injection under the auspices of their antigradation rules (Bloetscher et al. 2005).

Groundwater protection is achieved through rules that govern the construction and operation of injection wells in such a way that the injected fluid does not migrate into a USDW. The UIC rules regulate disposal of waste, particularly hazardous waste, via underground injection. The delineation of a USDW, as well as the requirements for construction wells to prevent migration, indicate that the regulations are focused on preventing injection of constituents that might impact USDWs.

Because the set of rules dealing with underground injection is lengthy, the intent of this chapter is to provide an overview of the various rules and their intent, with references to the complete federal rules and state rules for more information. The following topics are covered:

- Federal regulations
  - 40 CFR 144, 146, and 148 – Underground Injection Control
- Aquifer exemptions
- Zones of discharge

- Water rights and allocations by rule
- Wellhead protection
- State regulations for injection control
- Uses of waters of impaired quality

FEDERAL REGULATIONS – UIC

Federal regulations include an extensive set of definitions concerning the types and purposes of injection wells. Much of this oversight relates to the type of well proposed and its use. In addition, regulations have set forth types of materials that are permitted for use in well construction and minimum design criteria and testing to ensure well integrity. Federal regulations 40 CFR 144, 146, and 148 governing underground injection control (UIC) were established under the authority of SDWA approved in 1974 and amended in 1986 and 1996 (Tiemann 2010). These regulations segregate the wells into six classes, as described in Table 2-1 and illustrated in Figure 2-1.

Typically, Class V wells are shallow wells used to place a variety of fluids directly below the land surface. Examples of Class V wells are (from 40 CFR 146):

- Air conditioning return-flow wells used to return to the supply aquifer the water used for heating or cooling in a heat pump
- Cooling water return-flow wells used to inject water previously used for cooling
- Drainage wells used to drain surface fluids, primarily storm runoff, into a subsurface formation
- Dry wells used for the injection of wastes into a subsurface formation (injection of hazardous wastes into or above a USDW is prohibited by UIC rules)
- Recharge wells used to replenish the water in an aquifer
- Saltwater intrusion barrier wells used to inject water into a freshwater aquifer and create a subsurface density buffer zone to prevent the intrusion of saltwater into the freshwater
- Sand backfill and other backfill wells used to inject a mixture of water and sand, mill tailings, or other solids into mined-out portions of subsurface mines, whether the injectate contains radioactive waste or not
- Septic systems used to inject the waste or effluent from multiple dwellings and/or business establishments into a septic tank (UIC requirements do not apply to single family residential septic systems, nor to nonresidential septic system wells serving fewer than 20 persons a day used solely for the disposal of sanitary waste)
- Subsidence control wells (not used for the purpose of oil or natural gas production in those zones) used to inject fluids into a zone to reduce or eliminate subsidence associated with the overdraft of groundwater
- Injection wells associated with the recovery of geothermal energy for heating, aquaculture, and production of electric power
- Wells used for solution mining of conventional mines such as stope leaching
- Injection wells used for in situ recovery of lignite, coal, tar sands, and oil shale
- Wells used to inject spent brine into the same formation from which it was withdrawn after extraction of salts

Table 2-1 USEPA classification system for injection wells (Adapted from 40 CFR 144.6)

Injection Well Class	Uses
Class I	Class I injection wells are either wells used by generators of hazardous waste or owners and operators of hazardous waste management facilities that inject hazardous waste beneath the ground's surface. The waste is required to be injected beneath the lower-most formation of an underground drinking water source and more than 1/4 mile horizontally from it. This requirement includes other industrial and municipal disposal wells that inject fluids beneath the lower-most underground sources of drinking water (USDW). Class I wells can also be radioactive waste disposal wells.
Class II	Class II wells are used for the injection of fluids that are brought to the surface in connection with conventional oil and natural gas production, or the enhancement of recovery of oil and natural gas and the storage of hydrocarbons. A third subtype can only be used for liquid hydrocarbon storage; gaseous hydrocarbons are not included. At the state level, they are overseen by state oil and gas commissions and departments of natural resources. At the federal level, USEPA is the oversight agency.
Class III	Class III wells are utilized for the extraction of minerals.
Class IV	Class IV wells are for the injection of hazardous or radioactive materials into or above a USDW. Class IV wells are banned with the exception of wells associated with waste site cleanup under RCRA* or CERCLA† programs.
Class V	Class V wells are all wells that are not included in Class I, II, III, IV, or VI. The wells described herein for underground injection programs such as aquifer storage and recovery (ASR), saltwater intrusion barrier, or managed aquifer recharge, are all Class V wells.
Class VI	Class VI wells are used for sequestration of carbon dioxide.

* Resource Conservation and Recovery Act

† Comprehensive Environmental Response, Compensation, and Liability Act

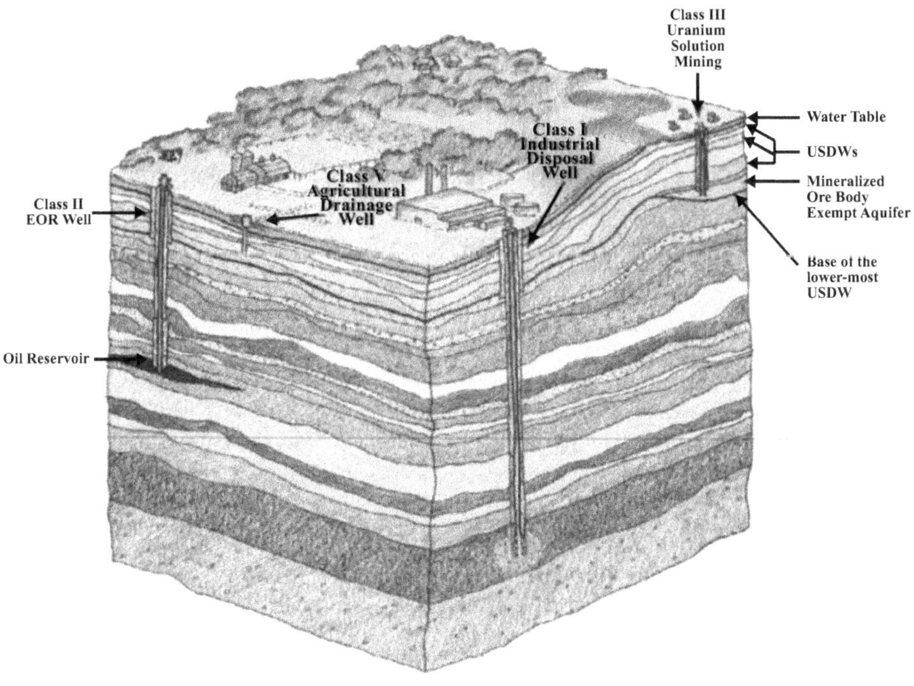

Figure 2-1 Different classes of wells

Source: *USEPA, http://www.epa.gov/region9/superfund/navajo-nation/pdf/stakeholders/UraniumPermits-DaveBasinger-USEPA.pdf*

The 1999 Class V Rule bans and restricts large-capacity cesspools and motor vehicle waste disposal wells.

States apply to USEPA to take on oversight and enforcement authorities (i.e., primacy) of the UIC program under SDWA Section 1422 (all well classes, with the exception of Florida where USEPA runs the Class II program) and SDWA Section 1425 (Class II for oil- and gas-related activities). 40 CFR 144.7 describes the means by which an aquifer may be designated as a USDW and refers to the specific USDW criteria presented in 40 CFR 144.3. The initial UIC regulations of 40 CFR 144.1(g) authorized the UIC Program Director to identify aquifers and portions of aquifers that are actual or potential sources of drinking water to help protect all USDWs. The regulations make clear that "[a]n aquifer is a USDW if it fits the definition, even if it has not been 'identified.'"

If an aquifer has not been specifically classified by the US Geological Survey (USGS) as a drinking water aquifer, it is assumed to be an underground source of drinking water. Narrative descriptions, illustrations, maps, or other geographic or geometric terms (such as vertical and lateral limits and gradient) are used to define the extent of USDWs.

AQUIFER EXEMPTIONS

Because the violation of any water quality standard may lead to the potential abandonment of a project, USEPA has established an aquifer exemption process. 40 CFR Section 144.7 establishes the process for state program exemption determinations. Aquifer exemptions are referred to as substantial and non-substantial (40 CFR 144.7(b)(3) and 40 CFR 145.32). USEPA may designate "exempted aquifers" using the criteria in 40 CFR 146.4. 40 CFR 144.1(g) as defined:

> [Exempted] aquifers are those which would otherwise qualify as "underground sources of drinking water" to be protected, but which have no real potential to be used as drinking water sources. Therefore, they are not USDWs. No aquifer is an "exempted aquifer" until it has been designated under the procedures in 144.7.

Otherwise, the SDWA does not mention aquifer exemptions explicitly, nor does the statute expressly preclude the exemption of an aquifer that is not currently used for drinking water purposes. The key provisions of the statute speak in terms of regulation, not absolute prohibition.

USEPA's aquifer exemption mechanism was promulgated in 1980 as a means of balancing underground disposal and mineral production needs with drinking water resource protection needs. USEPA believed that the most protective course was to adopt this broad definition and allow exemptions of aquifers from classification as USDWs on a case-by-case basis after review and public comment. This case-by-case approach was developed through the criteria for aquifer exemptions set forth in the regulations at 40 CFR 146.4 and 40 CFR 144.7 to provide the agency, after thorough site-specific analysis and public comment, a way to remove a USDW from UIC Program oversight.

An aquifer exemption is required if the water proposed for injection does not meet the water quality standards of the receiving water, based on its regulatory classification and a series of other conditions. In the final UIC regulations published on June 24, 1980, USEPA outlined four specific standards (40 CFR 146.4) to be met for the exemption of aquifers. The state is not required to designate an aquifer or portion of an aquifer as a USDW if:

1. It is not a current water supply, or cannot now or in the future serve as a water supply because it is mineral, oil, or geothermal energy producing

2. It is situated at depths or locations that make it economically and technologically impractical to produce drinking water
3. The aquifer is so contaminated that it would be economically or technologically impractical to render the water fit for consumption
4. In the case of a Class III well mining area, an exemption could be approved where the area would be subject to subsidence or catastrophic collapse in the absence of an exemption

Since the 1980 promulgation, the UIC regulations for exempted aquifers have been amended three times:

1. A modification of 40 CFR 146.4(c) to permit the UIC Program Director to exempt an aquifer not currently a source of drinking water if it contains between 3,000 and 10,000 mg/L TDS and "is not reasonably expected to supply a public water system"
2. A modification to the exemption provision in 40 CFR 144.7(b)(1) to allow the UIC Program Director to identify (by narrative description, illustrations, maps, or other means) and describe in geographic and/or geometric terms (such as vertical and lateral limits and gradient) all aquifers or parts thereof which the Director proposed to designate as exempted aquifers using the criteria in 40 CFR 146.4
3. A modification of 40 CFR 146.4 to state that the area covered by an aquifer exemption for a Class II enhanced oil recovery (EOR) or enhanced gas recovery (EGR) well may be expanded for Class VI injection wells for geologic sequestration, but that new aquifer exemptions for Class VI wells are not allowed

A substantial aquifer exemption is required if the proposed injection zone contains fluids with TDS of less than 3,000 mg/L. All substantial aquifer exemptions must be issued by USEPA in Washington, D.C., and a notice must be published in the *Federal Register*. Nonsubstantial requests may be issued by regional USEPA administrators.

Except in defining USDWs and aquifer categories, the federal regulations do not specify the technical criteria that are used by USEPA to judge aquifer exemption requests. As a result, USEPA developed additional criteria to evaluate each type of exemption request. USEPA will approve aquifer exemptions only for specific purposes. All exemption requests must include a description of injection activities allowed and a statement that additional approvals would be needed for other injection activities (e.g., hazardous waste disposal into an aquifer exempted for mineral production). Exemption requests must include economic criteria that address the availability of alternative supplies and their adequacy to meet present and future needs as well as the major costs for treatment and/or development associated with the use of the aquifer, including well construction and water conveyance (e.g., pumps and piping). The economic evaluation submitted by the applicant should consider these factors, as well as the following (Kimm 2000):

1. Distance from the proposed exempted aquifer to public water supplies
2. Current water supply sources for potential users of the proposed exempted aquifer
3. Availability and quality of alternative water supply sources
4. Analysis of future water supply needs within the general area
5. Depth of proposed exempted aquifer
6. Quality of the water in the proposed exempted aquifer

Approvals for aquifer exemptions can be difficult to acquire, especially for substantial exemptions. The most common reason for an exemption request occurs when the proposed injectate is of "lesser quality" because it contains one or more secondary drinking water quality parameters not found in the native water. Common constituents of concern with raw or reclaimed water are total organic carbon (TOC), fecal coliforms, disinfection by-products (DBPs), and nitrate. Microbiological constituents are not a major factor in the UIC program except for fecal coliforms, which are deemed to be an appropriate indicator organism.

Limited aquifer exemptions have been granted when the proposed injection zone contains TDS between 3,000 mg/L and 10,000 mg/L. Processing of applications for limited aquifer exemptions are a less lengthy process than for minor aquifer exemptions, and are oriented toward raw water ASR and reclaimed water applications. Rules designed for limited aquifer exemptions of alternative groundwater technologies primarily concern technical issues, where the original intent of the Safe Drinking Water Act is queried. A significant regulatory issue remains concerning whether the water quality standards should be measured at the wellhead or at the USDW (Bloetscher et al. 2005).

ZONES OF DISCHARGE

Occasionally, a state may permit the groundwater standards to be exceeded for some distance from the well. This is called a *zone of discharge* or *mixing zone*. A zone of discharge would be included in the permit to meet secondary criteria in the aquifer (Bloetscher et al. 2005). For example, Florida permits zones of discharge in G-II aquifers where the total dissolved solids is above 3,000 mg/L (Florida Administrative Code [FAC] sections 62-522.500 and 62-520). As applied, the injection system must be operated so the groundwater quality will be no worse than the natural background groundwater quality at the edge of the zone of discharge and the secondary sodium standard. In Florida, the extended zone of discharge can extend radially to the permit holder's property line, which may be greater than the 100 ft normally allowed for a zone of discharge (FAC 62-522.500 and 62-520).* The applicant may request an extended zone of discharge that extends beyond the property boundary at the time of initial permit application or with subsequent permit renewals or permit modifications, but the zones are not permitted to extend into a potable-well setback. The zone of discharge would generally be applied to parameters listed as secondary drinking water standards or for sodium. Zones of discharge are not generally provided for parameters listed as primary drinking water standards (except for sodium). Each state, where allowed, will define this differently, so local rules should be consulted.

WATER RIGHTS AND ALLOCATIONS BY RULE

The method of allocating water can be an issue in ASR programs from two vantage points: ownership of the injected water and rights to water in excess of permits or needs. In the eastern states, water allocations are typically administered under a riparian system of water law. Generally the water is owned by the state, not individuals, and water use above a specified amount may be required to be reported. In some eastern states, groundwater withdrawals may be regulated within specific aquifers. States where water withdrawals from designated aquifers are regulated include Georgia, South Carolina, North Carolina, Virginia, Maryland, Delaware, and Florida.

* Classification is specific to Florida; Class G-II waters are still potable, but have a total dissolved solids content up to 10,000 mg/L.

The picture is different in the western 18 states where water use rights are governed by the prior appropriation doctrine (PAD) of water law which, depending on the state regulations, may be applied. In most western states, the doctrine applies to surface waters and groundwater, but in Texas it applies to surface waters only. The PAD of water law establishes the right of a water user to use a specific amount of the state's water resources for a specific beneficial use or as associated with correlative land rights, such as in California, as stated in the water right by rule or statute. The water right holder has a priority date determined by a court or agency. The water users with older or more senior water rights have the right to use their water first. If the demand for water is greater than the available supply, the senior water rights holders can "place a call" on the water source (typically a river), which may result in junior water rights holders being prevented from using their water if doing so would impact senior water rights holders. Water users are restricted from changing the way they withdraw, divert the flows, or use their water to avoid a negative impact on existing water rights holders, including junior water rights holders.

For example, removing water from streams for placement in an ASR well is a water rights diversion that could conflict with the intent of western ASR rules. In addition, some states do not regulate groundwater except by regulations or special districts that restrict pumping levels to prevent land subsidence, so carefully laid plans might be frustrated by unregulated or unknown withdrawals by other entities. Little water law exists on how such migration of injected water would be evaluated with respect to water rights. Therefore, in a water rights state, the utility wanting to pursue an injection program may be required to acquire two water rights: one allowing it use of the source water injected and a second allowing it to withdraw that same water later. ASR facilities can apply for junior rights during times when excess water is available so they can withdraw the water from storage at times when water would typically only be available to senior water rights holders. The injection program also must demonstrate that the operations of the facility will not impact more senior water rights holders.

Water rights and permit issues change constantly with court cases and legislative actions. Refer to additional publications for more information on water rights. For example, Bernosky's *Overview of Environmental Laws and Regulations* (2011) provides an introduction to environmental laws that directly or indirectly affect water and wastewater utilities including selected state environmental regulations, riparian water rights, prior appropriation rights, and ordinances. Other resources can be found in the references section at the end of this chapter.

WELLHEAD AND SOURCE WATER PROTECTION

Wellhead protection is a potential regulatory concern for ASR programs. The protection of public water supply wells through wellhead protection programs is considered an important component of comprehensive state programs to protect groundwater from contamination. Many states have merged wellhead protection with source water protection. Other states are maintaining their wellhead protection program because of a specific federal grant fund.

Either way, the genesis of protection came in the 1986 amendments to SDWA, which increased the authority and responsibility for drinking water protection. Section 1428 of SDWA created the wellhead protection program (WHPP). Subsequent amendments in 1996 further strengthened drinking water protection by establishing the source water protection program (SWPP). Because of the duplicative nature of WHPP and SWPP, many states have replaced their WHPPs with the updated SWPPs or incorporated their WHPPs into their SWPPs. Wellhead protection areas (WHPAs) can encompass land areas with different criteria. For example, in many states protection can be enforced by a specific radius from the wellhead location to ensure runoff from surrounding areas does not introduce contaminants that can reach the groundwater via the well bores and poorly constructed wells. In other cases, WHPAs are recharge zones or zones of contribution (ZOC) for the underlying aquifers that supply water to the pumping wells. Depending on the hydrogeological parameters, a recharge zone or ZOC for a deep drinking water well may not be contiguous to the wellhead and must be located far from the withdrawal point.

Each state must prepare a WHPP and submit it to the USEPA for approval. Many states have promulgated minimal rules, which delegate implementation of the program to local or regional entities. Consult state and local authorities regarding specific wellhead protection efforts when developing an ASR program.

Delineation of a WHPA is typically done through numeric computer modeling of groundwater travel time and pollutant (solute) transport. Numeric models can be complex and cover large areas. Within those areas, local governments often prohibit certain land uses which may conflict with the local interpretation of private property rights. In some states, private property rights laws dictate that if an outside agency's activity on the property impacts more than 10 percent of the property, the offending agency must pay. This could be a significant impediment for utilities seeking source water protection within private property boundaries. Other wellfield protection ordinances prohibit the use and/or storage of certain hazardous materials in zones around wells. AWWA has addressed wellhead protection under emergency planning for utilities and source water protection in G300 Standard for Source Water Protection. The standard is designed to provide guidance to utilities in dealing with this issue. The companion guidance document, *Operational Guide to AWWA Standard G300, Source Water Protection*, and AWWA Manuals of Water Supply Practices M21, *Groundwater*, and M50, *Water Resources Planning*, also provide discussion on numeric models.

STATE REGULATIONS FOR INJECTION PROGRAMS

During the Reagan administration (1980–1988), the federal government delegated a series of programs for administration by the states. The UIC program was delegated to 33 states, partial primacy was given to seven states, and USEPA is currently running 10 state programs. There are additional programs in various US territories, Washington, D.C., and tribal nations. For the most part, the rules set forth standards for the federal UIC program that are mirrored in the states, territories, and tribes. Where specific state rules may differ somewhat from the federal requirements, they must be at least as stringent.

UIC rules focus on the permitting aspects of an injection project. Some states add the caveat that the injectate must remain in the injection zone, and that no unapproved interchange of water between aquifers occur. ASR programs may have other caveats to address, especially where waters of impaired quality—stormwater, reclaimed wastewater, industrial wastes, and others—may have more stringent requirements. Where the UIC program was delegated to the states, significant ongoing issues often require continued input by USEPA into the program.

Where the states are involved in this activity, the regulation may be separated into two regulatory programs or agencies: one for regulation of water resources and the other for regulation of groundwater protection. In states with UIC primacy, the UIC program is typically integrated into the state groundwater protection program. However, UIC authority can be split between the environmental protection programs and oil and gas programs, especially in the western states. For ASR wells in those circumstances, UIC programs are typically regulated by an environmental protection program, but, especially in the western states, these agencies do not handle water quantity and water rights issues, so their water resources or management programs also would be involved. In some states (e.g.,

Oregon and California), state agency programs are divided into regional offices and regulated by a regional representative. While 27 states have ASR projects, Figure 2-2 shows those states with ASR activity or have ASR-specific statutes or rules, those states with no rules (that may or may not include ASR wells), and states with no rules or ASR wells. California, Colorado, Montana, South Dakota, Indiana, and Florida have UIC programs that are regulated by a joint program between the state and USEPA. Arizona, Iowa, Minnesota, Michigan, Kentucky, Tennessee, Virginia, Pennsylvania, and New York have USEPA-regulated ASR programs. All of the remaining states have ASR programs that are regulated by the state.

Washington has at least three fully operational ASR facilities and allows them to operate under temporary permits while the regulations are under development. South Carolina, Texas, and Wisconsin have developed ASR-specific rules based on experience in the state. Other states regulate ASR projects under their existing regulatory programs, although statutes or rules may be on the books that refer to ASR. California only has rules governing ASR projects that use treated effluent. Arizona has a separate regulatory program governing treated effluent ASR projects in addition to its program for raw water ASR projects. Florida has one ASR system that proposes to use treated wastewater effluent, but reclaimed water is regulated in FAC Chapter 62-610. In Kansas, the regulations were developed initially by the state water rights agency. In New Mexico, the state legislature passed the law, and the state water rights agency adopted regulations in 2013.

Only Georgia prohibits ASR wells. In 2000, the Georgia Legislature placed a moratorium on the development of ASR projects until the state could complete a study of ASR's potential environmental effects because of issues related to the Savannah River. The moratorium expired in 2014 and the state legislature continues to hold hearings on the issue. The Georgia Environmental Finance Authority (GEFA) has proposed a contract to conduct an ASR demonstration project in the southwestern portion of the state. The ASR facility would store water from a shallow aquifer in a deep, confined aquifer. GEFA will recover the water to augment local streamflow during dry portions of the year. The legislature has not yet given the go ahead to the project.

The majority of ASR systems are sought for seasonal water supplies. Seasonally available water, whether due to climate (including natural weather phenomena such as

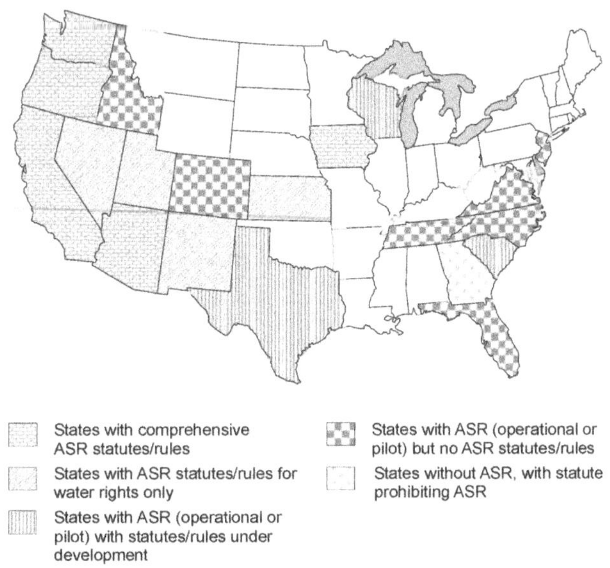

Figure 2-2 States with ASR-specific statutes or rules

Source: *AWWA 2002*

snow-melt) or periods of low demand, is used for injection periods, and the stored water is then recovered during another season with lower water availability and/or higher demands. Where long-term storage is expected, limitations on the amount of water that can be withdrawn after long-term storage may be set informally, based on ongoing monitoring by the ASR facility, or may formally be included in the agency permit provisions. For example, in Arizona, all ASR facilities using Central Arizona Project (CAP) water are subject to a 5 percent loss of storage per year as part of the state's efforts to recover groundwater levels in Arizona aquifers.

In some respects, the growth of rules has mimicked the growth in ASR sites nationally. Figure 2-3 shows ASR wells from the 1980s to date. As will be discussed in chapter 3, the 1990s and 2000s were active ASR periods in many states.

USE OF WATERS OF IMPAIRED QUALITY

Waters of impaired quality are nonpotable sources such as reclaimed water, stormwater, and raw water. There are no federal rules on the specific use of waters of impaired quality in Class V injection programs, so permitting of Class V wells is at the discretion of the UIC administrator (at the USEPA Region or state level). The federal regulations and many states have no specific permit requirements for ASR wells. As a result, state rules can allow the permitting and operation of Class V wells using reclaimed water as the injectate. For Class V wells, the required inventory requirement would allow the UIC administrator to determine if a specific Class V well may endanger USDWs and thus require it to be permitted. A water utility planning to develop an ASR well should discuss it with the primary agency to determine permit requirements.

ASR can be an effective and environmentally sound approach to the storage of reclaimed water for reuse systems, and would be considered a component of the overall

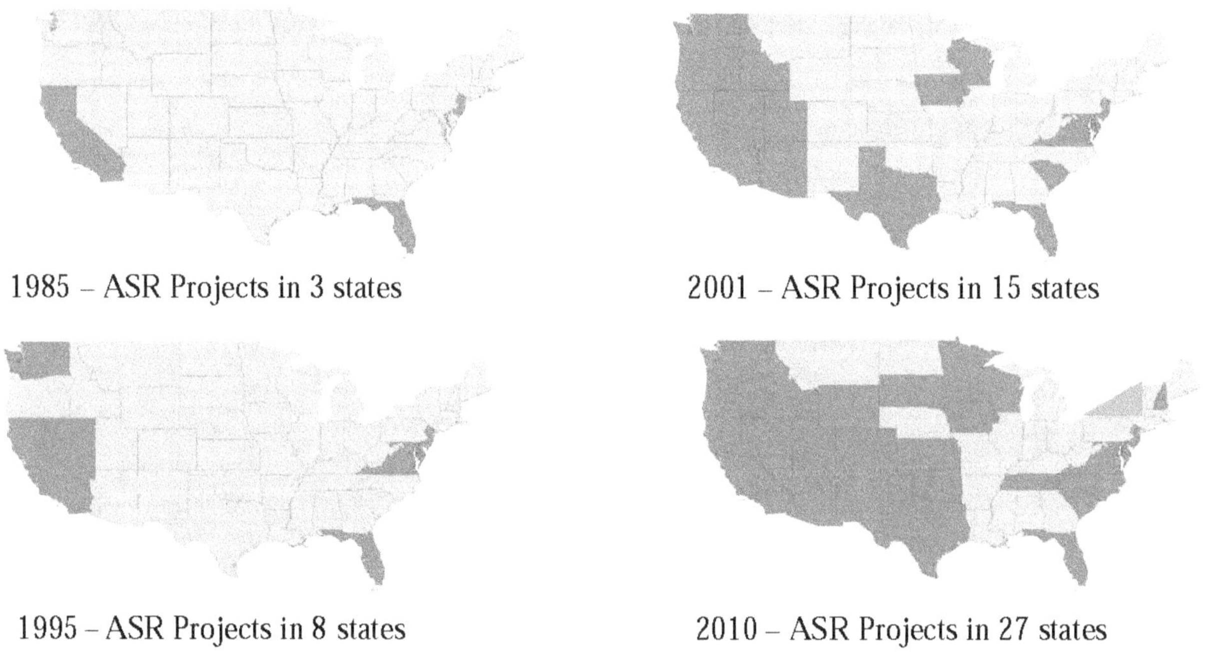

Figure 2-3 United States ASR projects from 1985–present

reuse system. Three states, California, Florida, and Arizona, specifically encourage the development of reclaimed water systems that integrate treated effluent as a part of state efforts to encourage water conservation and reuse. California has rules specifically addressing regulation of treated effluent. In Arizona, the extensive aquifer protection program encourages the reuse of effluent for nonpotable purposes, such as irrigating golf courses. Florida has one ASR system that proposes to use treated wastewater effluent (Englewood Water District, FAC 62-528).

The New Mexico Environment Department is currently exploring regulatory requirements for treated effluent ASR facilities with some water supply companies in the state. North Carolina and South Carolina regulatory staff have indicated that they would not allow ASR facilities that use treated effluent as the source water. Texas has one active reclaimed water ASR program.

ASR systems using reclaimed water must meet the technical and permitting requirements of the UIC program for Class V wells, including a construction and operation permit in addition to any permits required under the reuse rules. The water recovered from an ASR system typically must meet the quality standards for fecal coliforms before distribution.

Typically the reclaimed water recovered from the aquifer storage and recovery system will be monitored for total suspended solids (TSS), 5-day carbonaceous biological oxygen demand ($CBOD_5$), and fecal coliforms at the same frequency as a treatment facility providing reclaimed water to the reuse system. If reclaimed water withdrawn from an ASR system fails to meet one or more water quality standards, additional treatment, disinfection, and/or increased monitoring may be required to ensure compliance with these standards.

The distance between ASR wells using reclaimed water and potable water supply wells may pose a problem. The regulatory approach has been to require the ASR system to demonstrate that it will not adversely affect the quantity or quality of water withdrawn from the potable water supply well because confinement exists between the aquifer used for reclaimed water ASR and the potable water supply well. Some states require that the injection system be designed and operated to preclude movement of the injected reclaimed water to potable water supply wells.

Other injection programs plan for the use of reclaimed water to augment surface or groundwater resources used for public water supplies. Water quality in the receiving water and the classification of the aquifer drive the water quality of the injectate, as noted previously.

REFERENCES

AWWA (American Water Works Association). 2014. AWWA G300 Standard for Source Water Protection. Denver, CO: AWWA.

AWWA. 2014. Manual of Water Supply Practices M21, *Groundwater,* 4th ed. Denver, CO: AWWA.

AWWA. 2007. Manual of Water Supply Practices M50, *Water Resources Planning*, 2nd ed. Denver, CO: AWWA.

AWWA. 2002. *Survey and Analysis of Aquifer Storage and Recovery (ASR) Systems and Associated Regulatory Programs in the United States.* Denver, CO: AWWA.

Bernosky, Joseph J. 2011. *Overview of Environmental Laws and Regulations: Navigating the Green Maze*. Denver, CO: AWWA.

Bloetscher, Frederick. 2001. Risk assessment applications for alternative groundwater injection programs. ASCE-EWRI Annual Conference Proceedings – Orlando. Reston, VA: ASCE.

Bloetscher, Frederick, Albert Muniz, and Gerhardt M. Witt. 2005. *Groundwater Injection: Modeling, Risks, and Regulations*. New York: McGraw-Hill.

Bloetscher, F., A. Muniz, and A.M. Gonzalez. 2002. Summary of AWWA survey of ASR practices in the United States. Florida Section AWWA Annual Conference Proceedings, Hollywood, FL.

Englehardt, James D., Vincent P. Amy, Frederick Bloetscher, David A. Chin, Lora E. Fleming, Sinem Gokgoz, Helena Solo-Gabriele, Joan B. Rose, and George Tchobanoglous. 2001. *Comparative Assessment of Human and Ecological Impacts for Municipal Wastewater Disposal Methods in Southeast Florida: Deep Wells, Ocean Outfalls, and Canal Discharges*. Coral Gables, FL: University of Miami.

Florida Administrative Code. 1997. Chapter 62-610 and Chapter 62-528.

Kimm, Victor J. 2000. Guidance (#34) for Review and Approval of State Underground Injection Control (UIC) Programs and Revisions to Approved State Programs. Memorandum to Water Division Directors in Regions I–X. USEPA Office of Drinking Water (WH-550), Washington, DC.

Sham, C.H, R. Gullick, S. Long, and P. Kenel. 2010. *Operational Guide for AWWA Standard G300, Source Water Protection*. Denver, CO: AWWA.

Tiemann, M. 2010. Safe Drinking Water Act (SDWA): A Summary of the Act and Its Major Requirements. Congressional Research Service, CRS Report for Congress, December 2010.

US Environmental Protection Agency (USEPA). 1983. Underground Injection Control Program Rules. US Code of Federal Regulations › Title 40 › Chapter I › Subchapter D › Part 144 and Part 146.

USEPA. 2000. Ground Water Rule. http://water.epa.gov/lawsregs/rulesregs/sdwa/gwr/regulation.cfm.

USEPA. 1999. The Class V Underground Injection Control Study, *Volume 20—Saltwater Intrusion Barrier Wells*. Washington, DC: USEPA.

USEPA. 2002. Safe Drinking Water Act. www.epa.gov/safewater/sdwa/sdwa.html.

USEPA. 2002. Primary Drinking Water Standards. www.epa.gov/safewater/mcl.html.

USEPA. 2012 Guidance for the Determination of Underground Sources of Drinking Water (USDWs). Regional Guidance #3, last updated 6/14/2012. www.epa.gov/r5water/uic/r5guid/r5_03.htm.

Chapter 3

Summary of ASR Programs in the United States

This chapter summarizes the first complete study (conducted for this manual) of all of the known ASR sites in the United States, regardless of their planning, testing, or operating status. As of July 1, 2013, a total of 204 sites have been identified. The focus of the survey was on US ASR wells because Canada has only one (see Appendix B) and Mexico has none. Outside the United States, the regulations—where present—vary widely, as do construction methods. When an apples-to-apples comparison is not applicable, methods have not been discussed, although some of these methods are discussed in other literature.

ASR SITES IN THE UNITED STATES

ASR facilities are found in 27 states, most of which have fewer than five sites (many have only one or two). Figure 3-1 shows the number of sites in each state with ASR programs. The state with the most ASR programs, either completed or in process, is Florida, with more than 50, followed by California, New Jersey, Arizona, and Oregon. These five states account for the majority of ASR programs in the nation. Notably, most ASR programs in Oregon appear to be relatively new ventures, while many of the programs in Florida and California were initiated pre-2000. Figure 3-2 shows the growth of ASR installations since the 1960s by decade. There was a marked rise in the 1990s and early 2000s, but the number of new sites added appears to have decreased.

Figure 3-1 identifies the number of ASR sites by state but does not indicate the number of wells that are currently in use in each state. The 204 total sites include wells that

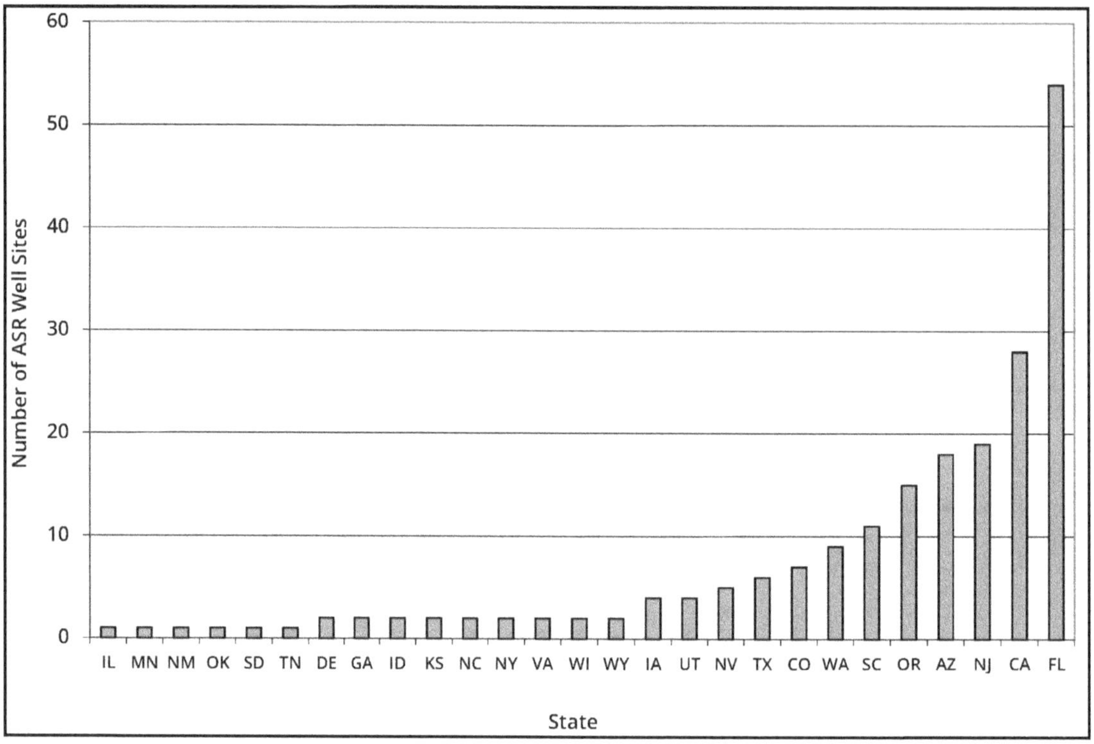

Figure 3-1 Number of ASR sites by state

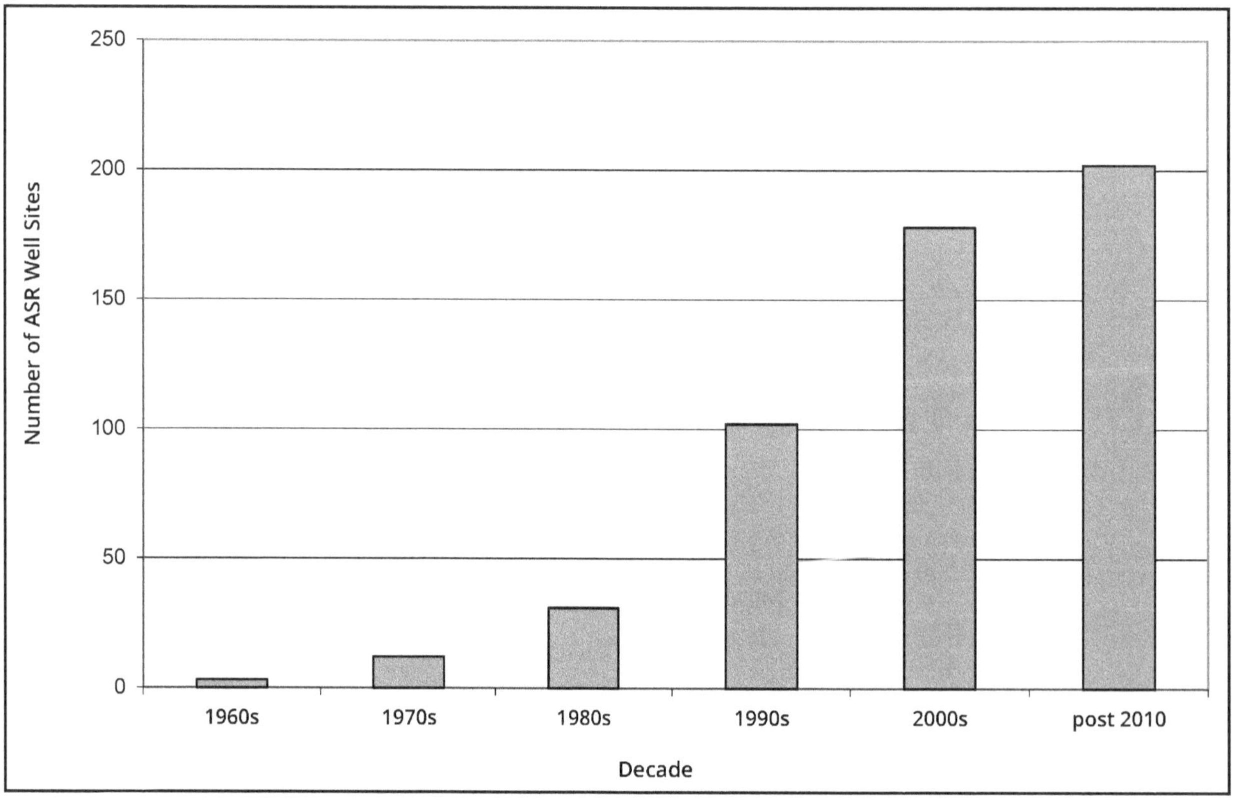

Figure 3-2 Cumulative ASR sites by decade

are no longer in service, were converted to other uses, were not pursued after testing, are dormant or inactive for a variety of reasons, or were plugged with concrete and formally abandoned. These wells are relevant because they point toward challenges encountered with the ASR concept as explored in chapter 4.

Water Use

Figure 3-3 shows the sources of water used for ASR programs, which tend to vary geographically. The largest source of injected water is surface water, followed by ground and reclaimed sources. In older projects, the ground and surface waters are normally treated to potable standards before injection. For example, Florida (groundwater) and South Carolina (surface water) have primarily focused on storing treated potable drinking water for later retrieval during peak-use seasons. In both cases most of the injected water is potable, having been treated in a water treatment facility. California has mostly used surface water flows as distributed by Metropolitan Water District of Southern California or that occur during the spring to be stored for potable water supplies during drought periods. Florida, Nevada, and Arizona use reclaimed water for irrigation purposes in some of their ASR programs.

The different uses of water recovered from ASR projects is depicted in Figure 3-4. The most common use of the recovered water is for potable purposes. Some of this water is retreated to address water quality changes introduced by the ground formation (such as metals, suspended or dissolved solids), but many potable supplies are simply disinfected and have stabilizing chemicals added. Raw water that needs to be treated is the second most common use of the recovered water.

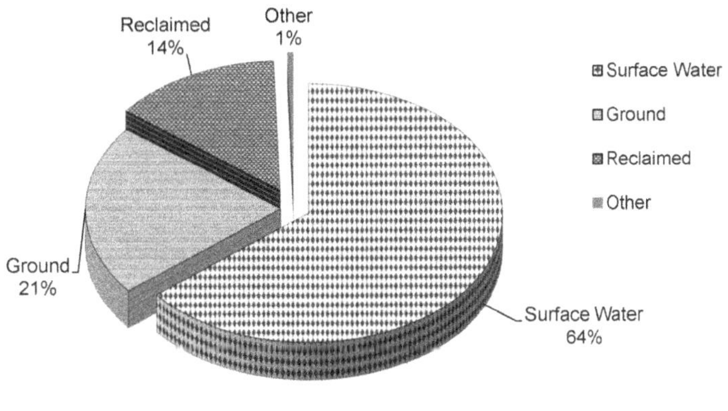

Figure 3-3 Sources of water used for ASR programs

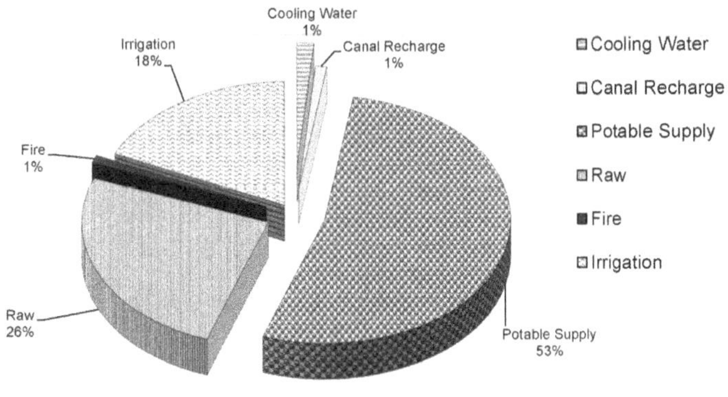

Figure 3-4 Uses of water recovered from ASR wells

Geology

ASR wells require the appropriate geology to function effectively. Important characteristics include rock matrix type and the hydraulic conductivity/transmissivity of the formation. These characteristics indicate how well water moves through the rock. Hydraulic conductivity (*K*, gpd/ft^3 or m/s) is a measure of the ease with which fluid is transported through a porous media, while transmissivity (*T*, gpd/ft^2 or m/day) is the aquifer characteristic that is defined by the rate of flow per unit width through the entire thickness of an aquifer per unit gradient. The transmissivity of the injection horizon (zone) varies widely, but is most commonly less than 10,000 gpd/ft^2. More than 80 percent of the ASR systems now in use have transmissivity under 100,000 gpd/ft^2.

A number of different formations are preferred for injection. The most common formations are limestone and alluvial (sand and gravel aquifers of glacial origin), although sand, sandstone, and basalt formations are also used (Figure 3-5). Although data are lacking for many wells, the confining unit information indicates clay, dolomite, and shale are common (Figure 3-6). A confining unit is a low-permeability geologic unit that forms either the upper or lower boundary of a groundwater flow system. Confinement, particularly vertical confinement, is critical to success as it prevents migration of the water out of the injection horizon. The injection zone is the geological formation, group of formations, or part of a formation receiving fluids through a well.

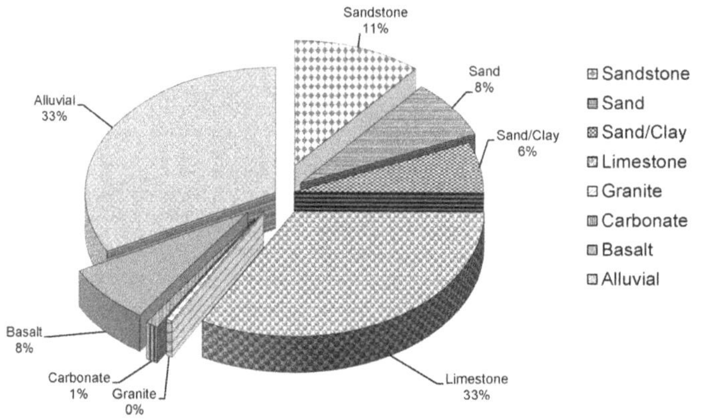

Figure 3-5 Injection horizon formation type

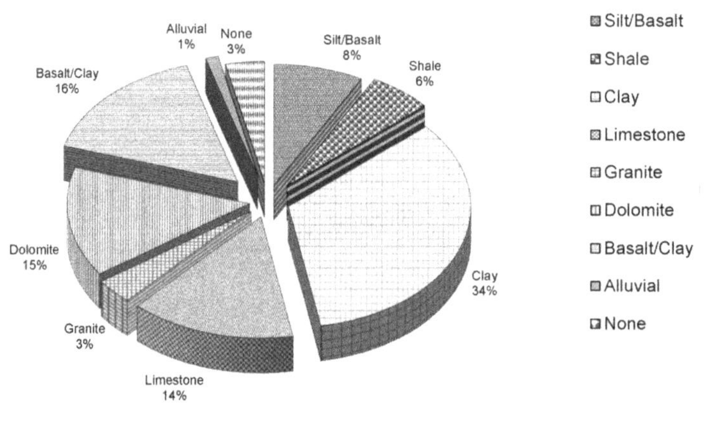

Figure 3-6 Confining unit formation type

INDIVIDUAL STATE ASR SYSTEMS

This section summarizes the ASR data collected from individual sites, by state for the period from June 30, 2012, to July 1, 2013. (When pursuing new projects, check with state regulatory agencies for more recent data.) The major issues each state faced when investigating the use of ASR are addressed to the extent possible.

Arizona

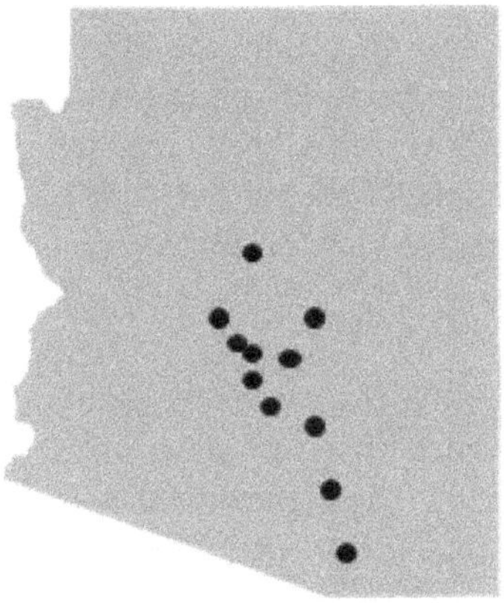

The state of Arizona has investigated the program at 18 sites. Fifty percent of Arizona's ASR projects inject raw water (i.e., Central Arizona Project [CAP] water) and the other half uses reclaimed water. Most of the projects were started in the late 1990s or early 2000s. All projects inject into alluvial sand/gravel formations, many of which are not confined. The systems are primarily in the Tucson and Phoenix areas.

Arizona has the greatest variation in construction; older wells typically are made of steel, while newer wells are constructed with PVC, stainless steel, and fiberglass casings. PVC is primarily a vadose zone (unsaturated portion of the aquifer) option. The typical well is 12 or 18 in. in diameter, but that can vary. The vadose zone wells are generally less than 200 ft deep because of large drawdowns of the aquifer from past use, but some ASR wells are deeper than 1,000 ft. Injection rates are generally under 1 mgd, although four projects were designed to inject more than 2 mgd. Withdrawal rates match or exceed the injection rates. Given the large number and age of the projects, most have had more than 10 injection/withdrawal cycles.

The oldest ASR system began in 1995 as a part of the Salt River replenishment project. Phoenix Water currently operates a large potable ASR project that has received several awards. The City of Chandler operates two ASR sites, including the largest operational system in Arizona with a total of 21 ASR wells at two sites: Tumbleweed and Ocatillo. Sun Lake City has five operational ASR wells and an additional two in development. Fountain Hills, Gilbert, Glendale, Honeywell, Peoria, Tuscon, and Gillette also have ASR projects. Of the 13 active ASR sites in Arizona, 10 use reclaimed water as their source for injectate.

California

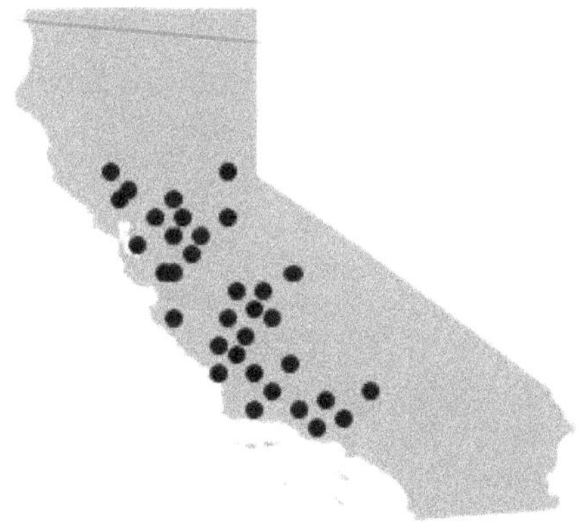

The state of California has 28 ASR systems. The largest and oldest is Calleguas (1993) with 18 wells, followed by Goleta with 12 wells. The California ASR projects are typically designed to capture excess surface water, much of which is treated prior to injection. However, in the ensuing years excess water was rarely available, so a number of projects have not injected in years. Most projects inject into alluvial sand and gravel formations. Some are confined by clay, but many do not appear to be confined.

The typical California well has a 16-in. diameter. Well depths typically are 300 ft to 500 ft deep because of large drawdowns of the aquifer from groundwater use in the past, but four wells exceed 1,000 ft. Most California wells are constructed with perforated steel. Victorville has an epoxy-coated stainless steel well. Injection rates are generally just under 1 mgd, although several were designed to take more than 2 mgd. Withdrawal rates range from equaling to doubling the injection rates. Fifteen projects were started in the mid-2000s, with the remaining initiated between 1990 and 2000. The number of injection cycles was difficult to determine as most are not reported. Given the limitations in the availability of excess water available from Metropolitan Water District of Southern California (the largest distributor of raw water in the state) because of drought and pricing considerations, many projects have had only limited injection. Two have converted ASR wells to raw water supply wells.

The most successful injection programs in the state, the Orange County and West Basin Groundwater Injection programs, were excluded from the survey and discussion in this manual because they do not recover water from the wells into which water is injected and thus are not true ASR systems. California has had success with ASR and continues to develop ASR projects in Santa Barbara, Foothills, Monk Hill, Long Beach, and Tracy.

Colorado

The Denver basin, which overlies four aquifers including the Arapahoe Aquifer, contains all but one Colorado ASR project. These projects inject potable water from surface water sources into sandstone or limestone formations. The typical well diameter is small (under 10 in). The Arapahoe Aquifer is far more favorable for accepting injected water than overlying formations, so the wells here tend to be deep—exceeding 650 ft for the casing and over 1,000 ft for the bottom of the well. The deepest reported ASR well is 1,538 ft, serving the East Cherry Creek Valley Water and Sanitation District.

The goal, in part, of the Colorado ASR projects is to help reverse impacts from large aquifer drawdowns from past groundwater use. Injection rates are generally under 1 mgd, with withdrawal rates about 50 percent higher. Most of the projects were started in the late 1990s and have had more than 10 injection/withdrawal cycles. The City of Centennial has continuously expanded its ASR wellfield since the 1990s, and now has 36 wells in operation. All the ASR wells in the state are steel except the newest Centennial wells, which are PVC.

Disinfection by-products were an emerging issue of ASR projects in the 1990s because of the injection of chlorinated potable water into formations that contained organics. In 1995, extensive work at the Willows Water District system (now part of the East Cherry Creek Valley Water and Sanitation District) demonstrated that trihalomethanes (THMs) dissipate in the formation with time. Since then, however, ASR exploration in the area has diminished considerably, primarily because of the lack of recovery and the ease in securing excess water in the 1990s. Douglas County, which is where the bulk of the Denver aquifer exists, is the fastest growing county in the state, and the ease in securing excess water from seasonal runoff and other sources is ending and ASR may become attractive again. Only the Centennial system is operating at more than 1 mgd, although the East Cherry Creek Valley Water and Sanitation District wells are operational.

Delaware

Delaware has two ASR projects. Both inject potable water into sand formations. The wells have a 12-in. diameter with an injection zone 250 ft to 560 ft deep. The wells are steel with stainless steel screens. Injection rates are generally under 0.4 mgd. Withdrawal rates match the injection rates. United Water's project, begun in 2008, is currently active. The Artisan project in Wilmington is listed as operational, but little information is available.

Florida

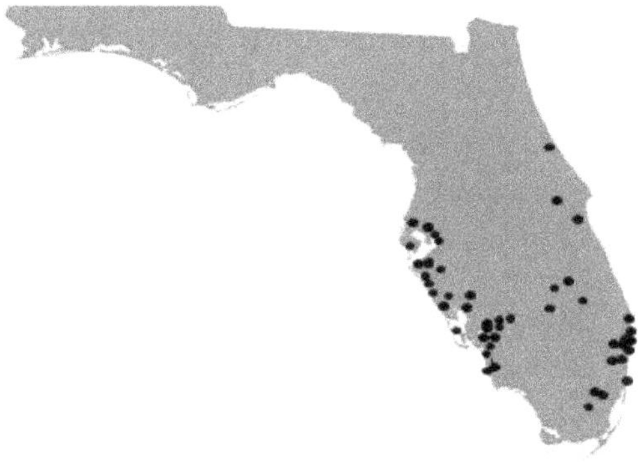

As noted, the state of Florida has more wells than any other state with installations at 54 sites. The Florida projects have injected raw, surface, ground, potable, and reclaimed waters, and as a result have the most variations of any state. Most of the wells were started in the 1980s and 1990s, but activity continues to this day. Virtually all projects inject into the limestone formations of the Upper Floridan Aquifer, although a couple of west coast systems use limestone seams in the Hawthorn clays as injection zones.

The typical well diameters are 12 in., 16 in., and 24 in., depending on the source water and goal of the project. The depths vary with the gradient of the injection zone formation, so the wells may be only 300 ft to 400 ft deep in central Florida, but more than 1,000 ft deep in southeast Florida. In theory, all the wells are confined below clays, although data from the Class I study (Gao 2010) indicates the confinement may be limited in central Florida. The state matches Arizona in terms of the variety of construction materials, although most Florida wells are made of steel with no screens. A smattering of PVC, epoxy-lined steel, and fiberglass wells exist. Injection rates vary greatly, but most injection is at a rate around 1 mgd. Withdrawal rates from ASR wells in fresh aquifers generally matches the injection rate; in brackish aquifers, the rate may be half the injection rate because of the migration upward of the freshwater in the formation. Successful long-term projects have had 20 or more cycles.

There are several studied, successful projects in Collier County, Cocoa Beach, Peace River, Rome Avenue, and Boynton Beach, among others, and a number have undergone expansions. Where issues about recovery, metals leaching, or other problems have arisen, the ASR wells generally have had less than 10 injection/recovery cycles.

Issues have arisen at a number of the 54 ASR sites in Florida; arsenic leaching and lack of recovery from brackish aquifers are the most common. Success varies considerably even among wells that are in close proximity. For example, the ASR well in Broward County was in close proximity to the successful Boynton Beach sites, but recovery from the Broward well was poor, and it was subsequently abandoned.

ASR expansion is on hold for a number of Florida systems because of arsenic leaching. A September 2013 memorandum from USEPA to the state attempts to address this issue. The memorandum applies only to potable ASR wells and permits the state to develop a mechanism to allow withdrawal of water with arsenic levels above the 0.01 mg/L standard if other tools are in place to limit public exposure. Those measures are left to the state to determine. The memorandum also notes that the utilities must maintain control of the aquifer to prevent unintended withdrawals by others. Exactly how the utilities can demonstrate control of water injected underground remains an ongoing discussion item with the permitting agency. The order may activate a number of stalled Florida ASR projects.

Georgia

As noted in chapter 2, Georgia is the only state that has had a law that specifically prohibits ASR wells. The law was allowed to expire in early 2014 as lobbyists worked the state legislature to lift the ban because the Georgia Environmental Finance Authority (GEFA) is proposing to conduct an ASR demonstration project in the southwestern portion of the state (Flint River). This ASR facility would transfer water from a shallow aquifer to a deep, confined aquifer for storage. GEFA would recover the water to augment local stream flow during dry months. The project is on hold until the legal issues are addressed. A potential challenge is that the formation proposed for injection is reportedly the same one that leaches arsenic in Florida. The ban has been reintroduced into the 2015 Georgia legislature.

Iowa

Iowa has the deepest ASR wells; all are over 1,500 ft deep. Four sites have installed ASR wells in Iowa. Two use treated groundwater for injection and two inject treated surface water. The two surface water sites have been in operation since the 1990s. The groundwater systems, Grimes and North Liberty, were established in 2012 and 2009, respectively. The receiving zone is sandstone with an overlying confinement of sedimentary rocks for all four sites.

The Grimes site is permitted for a well 2,110 to 2,602 ft deep in a sandstone aquifer. The injection rate is 1.1 mgd. The North Liberty site is proposed to inject into a horizon 1,575 to 1,673 ft below the surface into a sandstone aquifer. The capacity of the system is 1.1 mgd. Both wells are in the test stage.

Both surface water systems are in Des Moines and have one well each. Both wells are over 2,500 ft deep. The transmissivity is 10,000 to 30,000 gpd/ft. Both wells are 6 in. in diameter and made with steel casings. Injection rates are 1.0 and 1.5 mgd at the two sites, with recovery rate double that of the injection rate at both sites. More than five cycles have occurred at each site.

Idaho

Two sites have explored ASR in Idaho, and both have been in operation since the 1990s. One is an industrial facility (Micron Technologies) that uses the ASR-recovered water for cooling purposes. The other is a United Water–Boise project that was initiated in 1991.

The Micron Technologies project stores more than 8 billion gallons of water in three wells, which are drilled into a basalt formation. The transmissivity is 30,000 to 60,000 gpd/ft. All three wells are 8 in. in diameter and made with steel casings. Injection rates total 1.4 mgd with recovery rates of 2.1 mgd. Over 20 cycles have occurred at this site.

The United Water project was started in 1999 and includes four operating wells and one well that was deactivated due to clogging. The wells store treated surface water in a basalt formation with overlying basalt/clay layers; the injection zone is open from 70 to 220 ft below the surface. The wells are 10 in. in diameter. Only 15 million gallons of water have been stored in this project, although the capacity of the system is 1.1 mgd. Clogging is a reported issue with these wells.

Illinois

An ASR site was considered in Highland Park, Illinois, but it appears to have been inactivated before any construction occurred. There are no surviving details.

Kansas

Two sites have been investigated as potential ASR locations in Kansas, but wells have been constructed only for Wichita's Equus Beds project. The Equus Beds project includes four shallow vadose zone wells that store raw surface water in an alluvial formation in an injection zone open from 60 to 130 ft below the surface. The wells are 18 in. in diameter. Over 70 million gallons of water was stored. The injection capacity of the system is 10 mgd. The Republican River project was initiated in 2008 to investigate storing raw surface water, but no published data is readily currently available on its status.

Minnesota

The Joint Powers of Albertville utility has the only ASR well in Minnesota. It was constructed in 2007. It injects surface water into an alluvial aquifer 500 ft below the surface. The injection rate is 0.3 mgd, with recovery at 0.45 mgd. Limited information is available on the project, but it is operational.

New Jersey

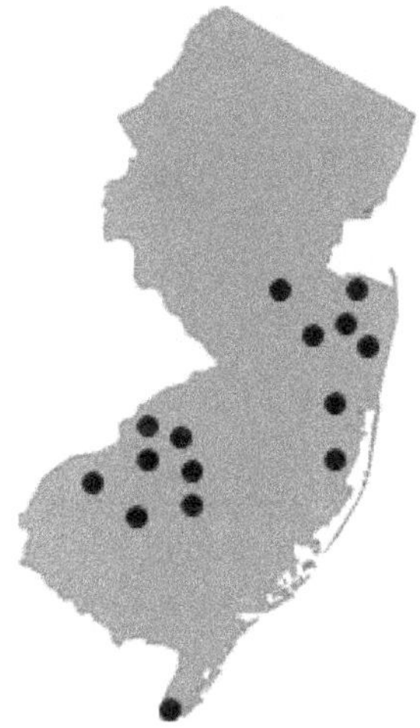

New Jersey is home to the United States' oldest ASR system; the Wildwood ASR system has been in use since 1968. Currently four operational ASR wells are located there. Since its inception in Wildwood, ASR has steadily grown and is considered a viable water storage option in the state. New Jersey has 19 ASR sites, with successfully operating wells in Gordon's Corner (2), Matchaponix (1), Brick Township (1), Evesham (1), and Mt. Laurel Municipal Utility Authority (1). An ASR project in Evesham Township is no longer active. Fifteen of these systems use treated surface water as the injection source. The remaining use treated groundwater. Additional projects are being developed in New Jersey, including projects in testing as of publication in Woolrich (Aqua New Jersey) and Swimming River and in development by Moorestown Department of Public Works.

The state requires a certain percentage of water to remain in the formation and the injection/withdrawal cycles are calculated on a calendar year. Most of the projects inject into a sand or sandy clay formation. The typical well is 12 in. in diameter; most are stepped down from 18 in., so 18×12 are common notations, while 24×16 and 20×14 exist. Some of these wells store as much as 2 billion gallons of water. The average depth is 550 ft for the casing and 690 ft for the well. The deepest system is 1,756 ft at Tom's River. All but one are steel, with many having stainless steel screens. Injection rates average 1.4 mgd, with withdrawal rates averaging 1.8 mgd. Most projects were started in the late 1990s. Given the large number of projects, and the age, most have had more than 10 injection/withdrawal cycles.

Nevada

Nevada has the most ASR wells of any state outside of Florida. There are five systems comprising 125 wells in Nevada, with the largest in Las Vegas Valley (LVV), which with 87 wells is also the largest ASR system in the country. The LVV project was initiated to offset prior aquifer depletion, but limited recharge water has been available in recent years due to other demands for reclaimed wastewater, which is the primary injection source.

North Las Vegas has 10 wells. The North Las Vegas and LVV projects were started in the late 1980s. The first Truckee Meadows project started in 1990. The remaining sites in the Truckee Meadows area near Reno came on line in the mid-2000s. A total of 28 wells have been constructed near Reno. The northern Nevada ASR projects use potable water derived from surface water for injectate. All projects inject into an alluvial sand and gravel formation.

The typical Nevada well diameter is 12 in., 16 in., or 18 in., but a variety of sizes exist. The depths range from vadose zone wells that are generally under 250 ft in Washoe County (the Truckee/Reno wells), to deeper wells, up to 1,200 ft in Las Vegas Valley. All but one well (Dayton) are steel; the outlier is PVC. Injection rates are generally under 1 mgd. Withdrawal rates match the injection rates except in the Las Vegas Valley, where withdrawl rates are lower than injection rates. The Las Vegas projects have had more than 10 injection/withdrawal cycles.

New Mexico

New Mexico has one inactive site, but significant work was done at this site. In Alamogordo, a study was conducted to evaluate historical spring flows and surface water supply that was available for injection into the La Luz wellfield aquifer and later recovered to augment the groundwater supply and provide for drought protection. One 18-in. steel-cased test well was constructed and tested. The project used potable water derived from groundwater as the injectate. The formation was limestone between 750 and 850 ft below land surface. Data at an injection rate of 0.8 mgd was collected for 30 days and a groundwater flow model was developed to simulate a full-scale ASR program. The study determined that ASR was feasible. No ASR well appears to be in place, however.

Utility managers in Rio Rancho and Albuquerque are looking at ASR in the future, but state regulatory staff did not identify either site as an ASR site.

New York

Two New York ASR projects were started more than 20 years ago on Long Island. The Bay Park investigation was conducted in 1968. The Bay Park project used potable water supplies derived from surface sources. Bay Park was a deep aquifer (418 ft to 480 ft) that clogged after 19 cycles and was subsequently discontinued. Injection rates were 0.3 mgd, while withdrawal rates were 0.6 mgd. The steel casing for the well had a 16-in. diameter. The formation was alluvial with a transmissivity of 1,000 gpd/ft^2.

The East Meadow ASR well used reclaimed water as the source that was injected into a surficial sand. The well had a 12-in.-diameter steel casing, open from 65 to 96 ft below land surface. Injection rates and withdrawal rates were 0.5 mgd. Fewer than five cycles occurred at this site. The transmissivity was 3,000 gpd/ft^2. This site is also no longer active.

North Carolina

North Carolina has two projects—one about 15 years old in Greenville and a new one that recently received a permit in Wilmington. Both projects inject potable water. About 15 years ago, Greenville was faced with a dewatered aquifer, requiring conversion to surface water supplies, and peak summer demands that required excess capacity in the system. The formerly dewatered Black Creek aquifer now acts as a reservoir for excess surface water that occurs seasonally.

The Greenville well, drilled in 1996, is a 12-in.-diameter steel well with open zones from 315 to 375 ft and 440 to 460 ft into a limestone formation. The formation has historically been productive. The well is injected at 0.4 mgd and recovers at a rate of 0.6 mgd. To date, 75 million gallons have been injected and stored. The well has been on hiatus for the last 10 years, but recent efforts to restart the well are underway. The success in South Carolina should bolster the prospects for this well because the Greenville well draws from the same aquifer (Black Creek) and suffers the consequences from the depletion of that aquifer.

The Wilmington wells are new. As of late 2014, one cycle test had been conducted in the first well and the second well was in testing. The injection rate is 1.3 mgd, with recovery at 1.9 mgd. These wells are 24-in. stainless-steel casings, with an injection horizon from 1,150 to 1,750 ft below the surface. The aquifer is listed as sand and gravel with a transmissivity of 10,000 gpd/ft.

Oklahoma

The US Geological Survey (USGS) conducted one test in Oklahoma in 1972 using an unrecorded number of now deactivated, steel-cased wells in Blaine County (the project is referenced as Blaine Gypsum in the literature). The project used surface water as the injectate. The formation was an alluvial sand and gravel formation less than 1,000 ft deep. All wells were steel. Injection rates were 1 mgd. Withdrawal rates were 50 percent higher than the injection rates. Twenty-five injection/withdrawal cycles were completed. This USGS demonstration project was decommissioned after no agricultural or utility entity was willing to continue operations and maintenance.

Oregon

The state of Oregon has 15 ASR sites. ASR began in Oregon in 1997 in Salem, which currently operates six ASR wells. Since 2000, the biggest increase in ASR activity in the United States has occurred in Oregon. Seven sites inject potable water, four inject raw surface water, and one injects filtered surface water. Three sites that inject raw water return the water for irrigation only, while another uses ASR for industrial purposes. Eight wells inject into basalt with overlying basalt/clay layers while the rest use alluvial formations.

The typical well has a 16-in. diameter, but sizes vary. The depths range from 200 ft in vadose zone wells used for irrigation purposes in the east to 600 ft deep in wells that are cased and have significant injection zones (one is 1,500 ft thick). The injection horizons are both basalt and alluvial formations. All wells are steel, and most have stainless steel screens. Injection rates average 1.1 mgd, although Pendleton was designed at 1.7 mgd. Withdrawal rates match or exceed the injection rates. Half the systems have had more than 10 injection/withdrawal cycles. Clogging was reported at several sites, including Pendleton and Portland.

South Carolina

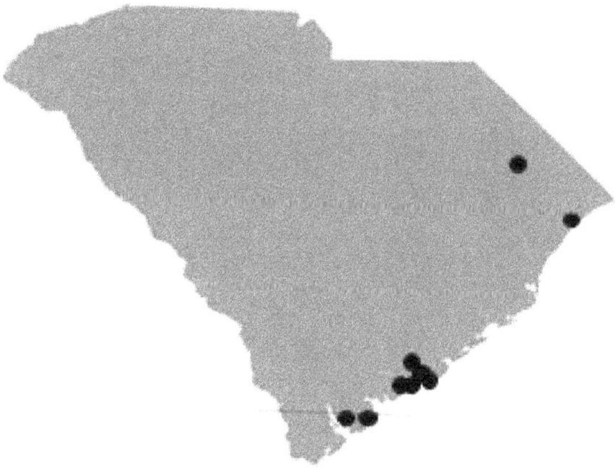

South Carolina has had major ASR success with 10 of 11 working systems. In the 1990s, South Carolina was faced with a dewatered aquifer, requiring conversion to surface water supplies, and peak summer demands that required excess capacity in the system. Implementing ASR took advantage of the dewatered aquifer and solved the lack of surface reservoir space and appeased high summer demands. The South Carolina systems now inject surface water into the Black Creek and Cape Fear aquifers, although the wells in the southernmost region are using the Middle Floridan Aquifer as a storage zone. Only the Myrtle Beach system was not pursued, although it was later merged with the Grand Strand system, which has been very successful in converting old potable supply wells to ASR wells. Beaufort–Jasper operates three ASR wells and Kiawah Island and Mt. Pleasant each operate two ASR wells. One inactive ASR well also exists in Mt. Pleasant. All but one of the projects (Orangeburg) injects into a limestone formation.

The limestone has a transmissivity less than 5,000 gpd/ft. The typical well diameter is 12 to 16 in. The casing depths range from 200 to 700 ft, with an average of 200 ft of injection horizon thickness. All wells are constructed with steel or stainless steel. Injection rates are generally under 1 mgd. Withdrawal rates match or exceed the injection rates. Given the large number and age of the projects, most have had more than 10 injection/withdrawal cycles.

South Dakota

USGS conducted one test in South Dakota using six now-deactivated wells. The project started in 1991 and used potable water originating from surface water as the injectate. The formation was an alluvial sand and gravel formation less than 60 ft deep. All wells were steel. Injection rates were generally under 0.1 mgd. Withdrawal rates doubled the injection rates. Five injection/withdrawal cycles were completed before the project was vacated.

Tennessee

Only one project exists in Tennessee. The project was constructed by Memphis Light Gas and Water (MLGW) and tested from 1996 to 2000. The source water was treated water from the Memphis municipal water supply that underwent aeration, filtration, and chemical addition. The well injected into a sand formation.

The well was 10 in. in diameter with a steel-casing depth of 346 ft and well depth of 394 ft. The well had a stainless steel screen. The injection rate was 0.8 mgd, and withdrawal 1.2 mgd. The well was put through five cycles of injection, storage, and recovery, before an issue with high trihalomethanes in the recovered water put an end to the program. MLGW's Water Quality Assurance Laboratory halted the ASR project because of the lack of indisputable evidence that no harmful by-products were produced as a result of the hydrogen peroxide pretreatment (see chapter 7).

Texas

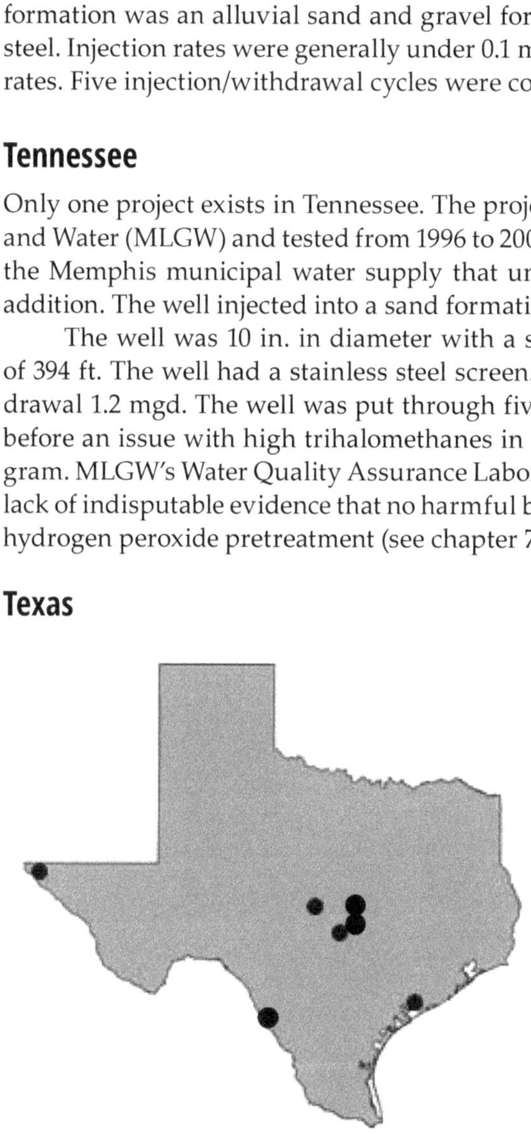

ASR began in Texas in 1985 in El Paso. The state now has three operational sites, one discontinued site, and two planning studies. El Paso uses reclaimed water as the injectate in its 11 operational wells. Kerrville injects surface water derived potable water, just as the now defunct project in Laredo did. San Antonio uses raw groundwater. All projects inject into sand or sandstone formations. The Texas State Water Board (TSWB) included ASR as a minor component (<1% of storage volume) of its 2012 plan, but recommended that ASR provide almost 81,000 acre-feet per year by 2060 (TSWB 2012).

Well data is lacking for the Texas sites. The only published diameter available is Laredo (6 in). The casing depths range from 280 ft deep to 620 ft. Well depths range up to 859 ft, with injection zones averaging 120 ft. All wells are steel. Except for El Paso (where data were not provided), all injection rates are under 1 mgd. Withdrawal rates match or exceed the injection rates. The three active systems (El Paso, Kerrville, and San Antonio) have had more than 10 injection/withdrawal cycles.

In 1990, Kerrville developed two ASR wells, and a third well is currently being developed. Kerrville recharges an aquifer with continuously declining static water level. The ASR injectate creates a temporary increase in aquifer pressure, but it is unclear as yet how much recovery is obtained. San Antonio is the largest system in Texas with 15 ASR wells. Corpus Christi is currently developing an ASR project.

Utah

Utah has four projects along with a host of other recharge programs. Two projects were started in the mid-1990s and two in the mid-2000s. The Jordan Valley Water District installed wells in the early 1990s, followed by Brigham City in 1998. Brigham City is touted by the state as having the largest successful ASR system in Utah. All projects inject into an alluvial sand and gravel formation except Leamington, which injects into limestone.

The typical well is 20 in. in diameter. The now-deactivated Salt Lake City project used vadose zone wells 15 to 40 ft below the surface. In the remaining active wells, screens start over 200 ft below the surface, and continue in depth to 1,000 ft below the surface. All casings are steel. Transmissivity varies widely—from 10,000 gpd/ft to over 300,000 gpd/ft. Injection rates are below 1 mgd/well, and recovery matches injection. Less than five cycles have occurred at three of the four sites. Clogging is a reported issue in Utah in alluvial formations.

Virginia

The two Virginia ASR projects were started more than 20 years ago. The Norfolk ASR investigation was conducted by the USGS in 1971, but the well was never converted to an operational ASR well. The Chesapeake system started in 1987. Norfolk used raw surface water, while the Chesapeake system uses potable water derived from a surface water source as the injectate. Both projects injected into a sand formation with a transmissivity between 5,000 and 10,000 gpd/ft.

The now-deactivated Norfolk well has an 8-in. diameter, with an injection horizon from 896 to 976 feet below land surface. A total of 20 million gallons was stored. Injection rates were 0.4 mgd, while withdrawal was 1.5 mgd. The Chesapeake system is still active, with more than 1 billion gallons of potable water stored in a 24-in.-diameter steel well, open from 560 to 745 ft below land surface. Injection rates are 0.1 mgd and the withdrawal rate is 0.5 mgd. More than 20 cycles have occurred at this site.

Washington

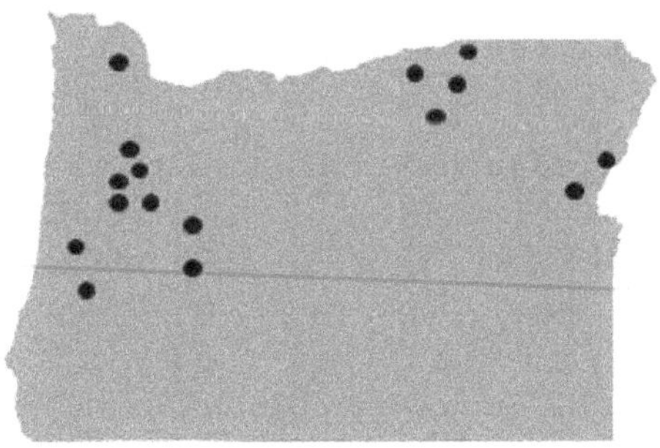

Washington has nine sites that have initiated ASR, starting with the Seattle Highline project in the early 1990s. The Washington ASR projects are raw surface water or potable water derived from surface water. Most of the projects inject into basalt overlain by basalt/clay confining units. The typical well is 12 in., but sizes vary across the state. The injection horizon depths begin at 200 ft at the casing base and are as deep as 959 ft. The injection zone is typically 50 or 250 ft thick. The wells are typically steel with stainless-steel screens. Most wells have had more than 10 cycles. Flow rates contrast sharply; the smallest are 0.2 or 0.3 mgd, while the largest range from 1.7 to 2.3 mgd. The larger capacity systems tend to use potable water instead of raw surface water supplies. Few transmissivity rates are known, but known rates are large: greater than 50,000 gpd/ft.

The potable supplies contain THMs, which are an issue with state regulators who are concerned about injecting water containing THMs into the formation, However, despite the regulatory issues, several more utilities are looking at ASR as a water supply option.

Wisconsin

The two Wisconsin ASR projects, in Green Bay and Oak Creek, were started concurrently in 1998. Both injected potable water derived from a surface water source into a sandstone formation. Both experienced metals-leaching issues.

The Green Bay well is steel-cased, 16 in. in diameter, with the injection horizon from 265 to 815 ft below land surface. Injection rates are 1.5 and 1.7 mgd for Green Bay and Oak Creek, respectively, with similar withdrawal rates. A total of 60 million gallons of water is stored at both sites. Less than 10 cycles occurred at either site. Transmissivity rates are 12,000 and 14,000 gpd/ft. Green Bay discontinued its ASR effort in 2003 because of high arsenic and manganese levels.

Oak Creek developed its ASR project in an effort to put off a treatment plant expansion. The Oak Creek wells have 19-in. steel casings, open from 584 to 1710 ft below land surface. In Oak Creek, the manganese levels in the recovered water were also too high, above the state's groundwater standards, but below the state drinking water standards. The administrative code in Wisconsin is very conservative, and the utilities did not wish to spend a lot of money treating the recovered water. The Oak Creek ASR project had not formally closed as of July 2013, when the Wisconsin Department of Natural Resources was waiting for a report from Oak Creek explaining what the city intends to do with the ASR well. Arsenic was reportedly also an issue with these wells.

Wyoming

In 2003, a pilot study was conducted in Gillette by the Wyoming Water Development Commission. Coal-bed-methane formation water was used as the injectate. Approval could not be gained to use any other type of source water. Because of the limited availability of coal-bed-methane water, the study was deemed not viable.

Currently, pilot studies are being conducted in Cheyenne for another project with an unspecified source water. The current well has a 10-in. casing, and is screened from 315 to 375 ft below land surface in a sandstone aquifer. The injectate flow rate is 0.2 mgd, as is the recovery.

REFERENCES

References for original data used for analyses in chapter 3 are found in the bibliography.

Gao, J. 2010. "Data collection and statistical analysis of deep injection wells in Florida." Thesis, Florida Atlantic University.

Texas State Water Board (TSWB). 2012. State Water Plan, 2012, Chapter 7, Water Management Strategies. www.twdb.state.tx.us/publications/state_water_plan/2012/07.pdf

Chapter 4

Challenges for ASR Programs in the United States

Subject to proper geologic conditions, ASR provides the potential for a viable method of improved water management. Based on data gathered from ASR wells, challenges may arise with ASR wells in four areas:

- Clogging of wells, which can mean physical chemical or biological fouling of the injection/intake interval and the receiving formation
- Inability to recover injected water in adequate percentages or volumes
- Water quality changes during storage (arsenic mobilization and disinfection by-products are the most common)
- Failure to meet expectations of clients; the reason some sites were abandoned or are now inactive (where a cause was not listed) may be unmet expectations

CLOGGING

Well clogging was reported as one of the major operational problems with injection-type wells, including ASR (USEPA 1999). Clogging was mentioned as a problem at more than 20 sites. There are three types of clogging: mechanical, chemical, and biological, as well as combinations thereof. Each will be discussed in this chapter.

Mechanical Clogging

Mechanical clogging problems result from suspended solids clogging the pores of the receiving formation, well screens, and/or gravel packs. Total suspended solids (TSS), turbidity, total dissolved solids (TDS), and entrained air in the recharge water are the principal physical characteristics of source water that play a part in physical clogging issues. The largest particles are suspended solids, which are the easiest to remove.

It is the rare ASR well that will not clog with high TSS (Pyne, Singer, and Miller 1996), which can be a problem with raw water supplies and reclaimed water. High TSS can be related or characterized by turbidity or silt density index (SDI). SDI represents the most influential factor in determining the potential for clogging, even when other factors may contribute to well clogging. Even some treated drinking water may contain high enough concentrations of materials to affect ASR well performance over time if the aquifer has low transmissivity. Despite the best efforts of operations, periodic clogging of the injection well is likely, as suspended solids and the chemistry of the injected water may reduce permeability of the near-bore formation surrounding the injection/intake interval of the well. The interstitial or pore spaces that are characteristic of porous media or unconsolidated sand aquifers are typically too small to permit movement of the suspended particles for any distance beyond the borehole.

If the well is heavily clogged by particles jammed into pores or cemented across screen slots, simple flow reversals may not prove sufficient to dislodge the particles. Such situations might be remediated by surging and pumping methods or by acidizing. Current water well practices to dislodge accumulating particulates include pressure fracturing and explosive fracturing, where the aquifer is broken apart by extreme physical or explosive force. However, hydraulic fracturing may not be permitted in every state (Pitt and Magenheimer 1997), so utilities should check with their state and local regulatory agencies before pursuing this action.

Brown et al. (2006) report that well clogging issues have been managed successfully through the use of a regular back-flushing programs. When the water flow is reversed by turning on the pump, the particles are removed from the well screen, gravel, and/or borehole wall, while the gravel pack is tightened. Several systems reported that redevelopment was required every three to five days to maintain well performance. Redevelopment of the wells often entails discontinuing recharge for a brief period and turning on the well pump to remove solids accumulating around the borehole. Notably, almost every well with a reported clogging problem used surface water as its source and injected into alluvial formations, indicating that the fines in the alluvia may have migrated under pressure from the well. Groundwater source systems also report clogging problems, however; Wildwood N.J., the oldest system in the United States, acidifies each of its four ASR wells every three months to combat clogging.

Determining the frequency of back flushing early in the project requires characterizing the TSS and turbidity of the recharge water through laboratory analysis and customized testing, such as the micro-fouling index test (AWWA 1991). Additionally, specific capacity and specific injectivity should be monitored to track trends and quantify the rate of decline. Back-flushing activities should include routine maintenance of the ASR well immediately after initiating testing or operational cycles (see chapter 6).

The lithology of the aquifer is another important factor when evaluating the cumulative potential for clogging. Thin layers of clay in the storage zone may swell and plug pore spaces in the aquifer adjacent to the clay beds. Clay minerals can also fragment during exchange reactions with recharge water. In aquifers composed primarily of clays and silts, plugging can result from both particle plugging and geochemical interactions between the injected water and the aquifer matrix or between the injected water and the groundwater.

The fragments migrate in the aquifer and reduce aquifer permeability, or they remain suspended, conveying elevated levels of turbidity in the recovered water.

In some cases, particulate plugging is caused by poor well design, including insufficient development of the well. Inadequate formation sampling may result in poor screen location. The rate of pumping is also important, as increased entrance and approach velocities can mobilize sand and silt particles within the formation but lying outside the well screen and gravel-pack filter. In some cases, the geophysical logging of the well prior to operation to evaluate water quality and formational characteristics may not be sensitive to thin layers of sand, silt, or colloidal matter that now may be exposed. Geophysical logging is a technique that uses probes lowered into the formation to retrieve electrical impulses that can be converted to useful aquifer data. In wells with gravel packs, incomplete development or over-pumping may be indicated by plugging of the gravel pack and the screens. Particulates also form a nucleus for chemical encrustation on a screen or column.

Pitt and Magenheimer (1997) suggest TSS concentrations as low as 0.5 mg/L to 1.0 mg/L can cause clogging of a porous-media formation, with significant loss in recharge rates of most wells with concentrations as little as 2.0 mg/L (Pyne 1995). Plugging caused by suspended solids can be mitigated by frequent redevelopment of the well.

TSS and turbidity are related, but differ in the method of measurement. Turbidity is often used as a substitute for the more tedious measurement of particles in water, and is quantified by measuring light scattered by particles in water without distinction to the types of particles in water. Thus turbidity provides no measure of health impact in itself. However, the particles associated with turbidity can interfere with the disinfection of water by shielding microbes from the disinfectant. Turbidity data, along with the microbial pathogen indicators total coliform and *E. coli*, provide an indication of the need for disinfection.

Mechanical clogging problems also include air entrainment (Pitt and Magenheimer, 1997), which can create air binding in wells. This occurs when cascading water in deep water tables with lower temperature water or high pressures forces oxygen out of solution. Other causes of air clogging may result in well damage, including loose piping connections, cavitation of the injection pump, inappropriate injection velocities, and source water quality. Pump cavitation should not occur during any operation and must be addressed immediately. Consequently, ASR wells should include a means for removing air from a well. Sealing the annulus—the void between the casing and the piping immediately surrounding—at the wellhead can prevent air entrainment during ASR operations (Pyne 2004). Air control can also be accomplished with a downhole control valve (Brown et al. 2006).

Cold water will have a higher viscosity than warm water, which adversely affects flow in the aquifer, with colder water resisting flowing into less permeable zones. The probability of clogging due to temperature differentials is site specific. The Colorado projects incurred this issue. Once the cold injectate displaces the native groundwater around the borehole during the first injection cycle, little temperature difference is found according to USEPA (1999).

Chemical Clogging

Chemical clogging has many potential causes, but most start with measurable total dissolved solids (TDS). TDS refers to the total inorganic and organic ionic material dissolved in water. Cations are positively charged ions in solution, and anions are negatively charged. Common cations found in water include calcium (Ca^{2+}), magnesium (Mg^{2+}), potassium (K^+), sodium (Na^+), and iron (Fe^{2+}). Common anions include chloride (Cl^-), bicarbonate (HCO_3^-), carbonate (CO_3^{-2}), and sulfate. Cations and anions indicate the level of, or potential for, certain kinds of pollution or harmful substances in water, and other characteristics such as hardness (AWWA 2014).

TDS cannot be seen by eye, but can be measured via electrical conductivity tests. Conductivity is a measure of the ability of a substance to conduct electricity and ionic activity. Conductivity provides an estimate of the TDS, which is comprised of the major cationic and anionic content of water, along with trace metals and nutrients. High TDS waters may indicate the potential to create physical clogging of the well via chemical means. Water with high conductivity generally indicates the presence of chlorides and metals in the water. High TDS is unlikely to create clogging problems, but may contribute to chemcial clogging problems by bringing ions into the formation that can create those chemical reactions. Corrosion is an example. Ongoing high TDS/conductivity injectate should be a consideration when screening recharge water quality.

Injected waters containing high sodium concentrations may exacerbate clogging by hydrating the clay, thereby causing the clay to swell. The injected water replaces stable native water, so the formation requires the water and aquifer matrix to reestablish equilibrium, which may take time and result in liberation of metals. Silt particles in alluvial formations may be disturbed along with clay colloids and clog the pores of the injection zone. To assess underground conditions, it is necessary to obtain accurate geophysical logs along with cores.

Native groundwater exhibiting a sodium-bicarbonate chemistry often indicates the presence of unstable clay minerals in the aquifer matrix. Often conventional wells screened in aquifers containing a sodium bicarbonate display difficulties with initial well development and turbidity. Areas where conventional production wells experience turbidity problems during development should be avoided for the siting of ASR facilities. Distinct areas of the Atlantic Coastal Plain in Virginia, Delaware, New Jersey and New York contain aquifers exhibiting a sodium-bicarbonate chemistry (AWWA 2014).

Carbonate precipitation can be controlled by lowering the pH of injection water. Existing problems can be solved by acidizing the well with a corrosion inhibitor that dissolves the precipitates. Steel and stainless steel are particularly subject to precipitates. Injecting carbon dioxide into the well screen can also remediate clogging.

Chemical-induced clogging is not unusual and can be treated with acidization of the well and redevelopment. See chapter 6 in this manual and M21, *Groundwater*, for more information on resolving chemical clogging problems.

Biological Clogging/Fouling

Biological issues with ASR wells are no different than in water supply wells, although the use of chlorine in injected potable water may help keep the biological issues at bay. Microbiological fouling is generally interrelated with physical and chemical processes. Microorganisms can encrust or corrode the system, enhancing physical and chemical deterioration. Well deterioration problems involve some microbiological fouling. The typical symptoms of microbiological fouling problems include the following:

- Decrease in the water quality
- Increased drawdowns
- Reduced specific capacity
- Change in the amount of iron or manganese in the water supply
- Increased microbiological densities, such as slimes or staining from the raw water (AWWA 2014)

Microbiological fouling encourages changes in the electrical potential and pattern of the well surface by creating anodes on metallic surfaces to transfer ions. Biofilms are formed that eventually clog screens. The bacteria absorb nutrients and minerals such as iron, manganese, arsenic, nitrogen, and oxygen within the biofilm matrix, leading to the formation of tubercles and films that reduce capacity of the pumps and casings. Wells with water-lubricated vertical turbine pumps can be subject to severe microbial action (AWWA 2014).

Aquifers are ideal environments for many microflora adapted for these conditions (AWWA 2014). There is tremendous surface area for colonization, the temperatures are relatively constant and moderate, the flow of water provides a consistent nutrient supply, and, except for the immediate pumping zone, the water is not disturbed. The cycling of a pump on and off with the exchange of atmospheric air in the casing provides the oxygen supply to maintain growth of certain bacteria in the well. All spaces within the formation are potential areas for colonization. Vugular formations (rock with open pockets) or those with significant fracture surfaces are ideal for creating large biofilms within the aquifer, but never demonstrate severe plugging because of the size of the organisms compared to the opening aperture size (AWWA 2014). A biofilm is an entire active ecosystem, providing an environment for survival to a variety of microorganisms by storing and transporting nutrients. The biofilm protects the bacteria cells from external reagents such as chlorine but traps iron, sulfur, manganese, and other nutrients. Precipitates of the iron, sulfur, arsenic, and manganese occur within the biofilm (AWWA 2014). The amounts of nutrients, carbon, and oxygen in the recharge water can greatly affect the potential for biological growth in an ASR well, and excessive biological growth can cause well capacity to decline. The most common form of biological growth is iron bacteria, but sulfate bacteria have also been observed in ASR wells (Pitt and Magenheimer 1997). Although the amount of nutrients in treated water is low, suspended solids that tend to concentrate near the well as a result of filtration will form a significant substrate for biological activity (Pyne 1995).

Biological growth can be controlled by maintaining a chlorine residual in the injected water. Note that a chlorine residual may create a trihalomethane issue—a problem in several states—which is discussed later in this chapter. Since most ASR systems use potable water or reclaimed water, a chlorine residual is already present in the treated water. If chemicals are added in the treatment process that may serve as a source of nutrients, alternative chemicals should be substituted that do not encourage biological activity.

WATER QUALITY

A second major problem observed at ASR projects is various water quality issues. The principal chemical characteristics of source water include pH, alkalinity and acidity, dissolved cations and anions, hardness, conductivity, and carbon dioxide. The chemistry of the injection water is an important consideration when determining the applicability of ASR for a given aquifer. Depending on the composition of the injection water, unwanted geochemical reactions may occur between the injection water and the native groundwater and/or aquifer matrix. Such reactions could lead to well clogging or degradation in quality of the stored—and ultimately recovered—water. Before implementation of an ASR project, the recharge water is usually analyzed to predict any future problems and to identify appropriate preventive measures such as treatment before injection, treatment of the recovered water, or special operating procedures. Generally, the lower the quality of the injection water, the higher the likelihood that it will cause problems with the ASR system. Specific elements or chemicals that affect water quality include nitrates and ammonia, sodium, phosphate, chloride, sulfates, iron and manganese, pesticides, solvents, and radioactive substances.

Leaching

ASR projects in several states reported issues with water quality changes, particularly metals mobilization. Several operational sites are currently dealing with arsenic (As), radon, and uranium issues. The arsenic issue is most commonly identified in Florida. With respect to the Florida arsenic issues, Arthur et al. (2002) reported that

> "examples of natural As mobilization mechanisms and associated mineral phases include: 1) oxidation of sulfide minerals such as pyrite, which may contain trace elements as lattice substitutes, impurities or in solid solution (Nickel, Cobalt (Co), Copper (Cu), Lead (Pb), Arsenic (As), Zinc (Zn) and Manganese (Mn)), 2) desorption or dissolution of iron (Fe) and Mn hydroxides (including grain or fracture coatings), 3) oxidation-reduction of organic material, which can mobilize organically complex As, and 4) biological transformations."

Arthur's study suggests that

> "from the whole rock geochemistry data, many trace metals such as As, chromium (Cr), Ni, Vanadium (V), Molybdenum (Mo), and Uranium (U) showed high concentrations, when compared to global averages for limestones. Linear covariations of As and Mo suggest that Mo is possibly associated in As-bearing phases such as arsenian pyrite or organic material. Other minerals or phases containing As in these carbonate rocks include organic material and Fe- and Mn-oxyhydroxide coatings."

The results of Arthur et al.'s investigation indicated that the geologists need to be concerned with chemical (including isotopic) variability within groundwater and the aquifer system. In particular, the injection of oxygen-rich surface waters caused the release of trace metals such as arsenic (As), iron (Fe), manganese (Mn), uranium (U), and perhaps nickel (Ni) into the stored water. The study's recommendations involved designing recharge–storage–recovery cycle tests and the location of monitor wells to better understand important geochemical processes during the storage of oxygenated recharge water (Arthur et al. 2002, Mirecki 1998). Other studies document arsenic mobility during artificial recharge of aquifers (Stuyfzand 1998; de Ruiter and Stuyfzand 1998, Brun et al. 1998), as well as mobilization of As, Co, and Ni as a result of oxidation (Stuyfzand 1998). Stuyfzand found that arsenic remains relatively mobile while nickel and cobalt are less mobile and likely coprecipitate or adsorb onto iron hydroxides further away from the injection well.

Pyne (2004) suggests pH differences between the recharge and groundwater can cause reactions that lead to clogging problems or to the release of constituents from the aquifer that may require removal upon recovery of the water. Closely related to pH are the characteristics of alkalinity and acidity. Alkalinity measures water's ability to neutralize acids, that is, its ability to react with H^+ ions. Acidity measures water's ability to neutralize bases, or its ability to react with OH^- ions. Chemical leaching issues must be solved with treatment (often in and out).

The dissolved oxygen content of surface waters is higher than that of native groundwater. The native groundwater can become oxidized once in contact with oxygenated or disinfected injectate. Elevated concentrations of recovered manganese and iron can occur when oxygenated recharge reacts with reduced metal bearing minerals in the aquifer like pyrite and the iron carbonate mineral, siderite. ASR wells that use surface waters as injectate are more vulnerable to this type of clogging than groundwater sources. Most groundwaters have a pH above 7 and significant dissolved solids. Surface waters contain less TDS, but more TSS. The pH of untreated surface waters is often lower than that of native groundwaters. The differences lend themselves to chemical leaching issues. In a New Jersey ASR site, oxidation of minerals within the aquifer by oxygenated recharge water has caused the iron concentrations in the recovered water to increase considerably above the secondary drinking water standard of 0.3 mg/L (Pyne 2004). When acidification with deoxygenated water failed, the iron concentration was reduced in the recovered water by elevating the pH of the recharge water (Lucas and McGill 1997).

Geochemistry is a difficult topic and involves many variables. For successful ASR operation, the injection water must be compatible with the native groundwater and the aquifer material. Water quality issues normally result from adverse reactions occurring among the quality of the injectate, the native groundwater quality, formation mineralogy, and subsurface temperature, as well as the pressure that occurs during injection (Pyne et al. 1996). Understanding the properties of the recharge water, native groundwater, and mineralogy of the aquifer can help predict potential adverse geochemical reactions and resulting changes in water quality. This approach allows operations personnel to adjust recharge water quality to minimize adverse reactions in the storage aquifer.

The solution to leaching is a challenge. It means that treatment of the water for stabilization with the native formation is required prior to injection. Leaching may suggest treatment of the recovered water is required as well.

Disinfection By-Products

Elevated concentrations of disinfection by-products (DBPs) in recovered water is an important issue that influences ASR programs in some states, particularly on the West Coast. Most potable and reclaimed water sources contain chlorine as a disinfectant. In the presence of organics in the treated water, haloacetic acids (HAAs) and trihalomethanes (THMs) may occur. Both HAAs and THMs may be considered contaminants in the injection formation, creating a regulatory issue. This issue has halted several projects in Washington and California. The issue was also raised in Florida.

The impact of THMs and HAAs on ASR projects in Denver, Colorado (Willows Water District), Las Vegas, Nevada and Lancaster, California, have been studied in detail (Pyne et al. 1996, Katzer and Brothers 1989, Baqai 2002). These studies and others demonstrate that HAA and total THM concentrations decrease during storage (Bouwer and Wright 1987, Landmeyer et al. 2000). By calculating the first-order attenuation rates from the rate of change in the mixing-corrected DBP concentration over time, Singer et al. (1993) and Pyne et al. (1996) suggest that THMs and HAAs are reduced during storage in the aquifer, possibly as a result of biological processes under anaerobic and anoxic conditions. Pavelic et al. (2007) report "three orders of magnitude difference in total THM attenuation rates occurred between the smallest half-lives at Charleston (< 1 day) and the greatest half-lives at East Bay (120–550 days) and Las Vegas (>120 days)." Pavelic et al. also note that degradation rates were found to be strongly dependent on the redox state. Sites that were aerobic had half-lives in excess of 50 days, those that were anoxic (nitrate-reducing) had half-lives of less than 70 days, and those that were anaerobic (sulfate-reducing) had half-lives of less than 20 days. Evidence for the continued formation of DBPs in aquifers during ASR is strong given current knowledge of DBP processes, but there has been little evidence apart from studies by Singer et al. (1993) and Fram et al. (2003) to conclusively demonstrate actual concentration increases and the kinetics involved.

Miller et al. (1993) and Bernholtz et al. (1995) suggest that the THM removals observed in the Las Vegas ASR system are the result of dilution with groundwater only, and the Foothill Municipal Water District in California has observed an increase in THMs, suggesting the formation of THMs in the subsurface environment (Palmer et al. 1997), likely due to organics in the injection horizon. Yet in both cases the THMs attenuate (decrease with time) after initially increasing. As a result, it is unclear whether DBPs are a long-term issue with ASR, but they certainly are a regulatory issue in some states. Utilities will need to work with regulators during demonstration projects to fully understand and assess the role of ASR and its potential, unique challenges.

Carbon Dioxide

Carbon dioxide is an end product of biological decomposition. High concentrations of carbon dioxide indicate biological activity in the aquifer and an acidic pH. Carbon dioxide was suggested as the issue with the Myrtle Beach well by a former director as the reason this program was discontinued. Since carbon dioxide is often associated with microbial activity, it might have been resolved with additional well development, which would have released the carbon dioxide from the formation over time.

LOW RECOVERY AND EXPECTATIONS

The inability to recover sufficient quantities of injected water was listed as a reason to abandon ASR efforts at 20 sites. This may in part be because of utility expectations; clients may expect that all water injected will be recovered. But sufficient water recovery may also be a challenge because of the tightness of the formation or the water quality of the receiving formation. To determine water quality of the recovered water and the amount of water that might be successfully recovered, a series of injection and recovery cycles must be undertaken. So as not to waste water, these injection and recovery cycles usually start small (less than 10 million gallons) and increase to 30–50 million gallons. Many systems have more than 100 million gallons invested in injection cycles; since most ASR wells have a 1 mgd capacity, this means 100 days of injection. Recovery may be more or less than the 1 mgd; it is one of the unknown variables of ASR (Bloetscher et al. 2003; AWWA 2014). A five-cycle test program might take two or three years on top of the permit and construction time. The client expectation that a well will be operational within a year or less is rarely attainable. In addition, the 100 million gallons may be lost if recovery is unfeasible because of the formation or the native water quality.

For success, confinement immediately above the injection horizon is necessary so the buoyancy of the injected water does not cause it to move away from the ASR well. Temperature also significantly impacts the density stratification of water, so the impact of temperature of the injection horizon must be evaluated. This means obtaining statigraphic information—something not done with all ASR wells—as well as water quality data. The key is finding the most appropriate ASR injection zone.

The use of ASR technology in brackish aquifer systems presents challenges related to the mixing of the injected "fresh" water with the native brackish water. Where a brackish aquifer is used, the density difference between the treated freshwater and saline sources causes a bubble of freshwater to form. A large mixing area may also result, depending on the formation, but with adequate quantities of injected water and the right aquifer properties, the bubble will remain fresh for a sufficient period of time to be available for water supply augmentation. The quantity of water to be stored can be as extensive as the localized aquifer system and can be pumped into the aquifer continuously until needed. However, the initially mixed water will not be recoverable without treatment to remove impurities such as elevated levels of chlorides (salinity). In many cases the transition zone may be over 500 million gallons for a single well (more in an ASR wellfield), after which the ASR system may recover 70 percent of the injected water, depending on aquifer characteristics. A density differential between the fresh and brackish water and the degree of confinement are critical to creating a bubble of freshwater within the aquifer zone (the density of fresh water is 1.00 g/cm^3 at 25°C, versus a seawater density of 1.025 g/cm^3).

For brackish ASR systems, optimization of the withdrawal rate may significantly affect recovery.

Ultimately the question for any ASR program is what percent recovery is expected or accessible. That answer could be 40, 70, or more than 90 percent depending on costs and utility needs.

REFERENCES

Arthur, J. D. 2004. Water-rock interactions during ASR and effects on water quality. *Aquifer Storage and Recovery.* IVAGWT/HC/FGS. April 15–16, 2004, Tampa, Florida.

Arthur, J.D., J.B. Cowartand, and A. Dabous. 2002. Mobilization of arsenic and other trace elements during aquifer storage and recovery, southwest Florida. US Geological Survey, Artificial Recharge Workshop Proceedings. April 2–4, 2002, Sacramento, California. http://water.usgs.gov/ogw/pubs/ofr0289/jda_mobilization.htm.

AWWA (American Water Works Association). 2014. Manual of Water Supply Practice M21, *Groundwater* 4th ed. Denver, CO: AWWA.

Baqai, Hisam A. 2002. Effects of recharge of chlorinated state water project waters to groundwaters in Lancaster area of California. *Water Engineering and Management,* 149(4):10–13.

Bernholtz, A., K. Brothers, and T. Katzer. 1995. Artificial groundwater recharge in the Las Vegas Valley, Clark County, Nevada: Storing today, treating tomorrow? Proceedings of the Second International Symposium on Artificial Recharge of Ground Water. American Society of Civil Engineers. July 17–22, 1994, Orlando, Florida.

Bloetscher, Frederick, and A. Muniz. 2001. Does the current knowledge of ASR answer all the necessary questions? Groundwater Protection Council Annual Conference Proceedings. Reno, Nevada. Oklahoma City, OK: Ground Water Protection Council.

Bouwer, E.J., and P.L. Wright. 1987. Transformation of trace halogenated aliphatics in anoxic biofilm columns. *Journal of Contaminant Hydrology,* 2:155–160.

Brown, Christopher, J. Kirk Hatfield, and Mark Newman. 2006. *Lessons Learned from a Review of 50 ASR Projects from the United States, England, Australia, India, and Africa.* Jacksonville, FL: University of North Florida.

Brun, A., F.D. Christensen, J.S. Christiansen, P.J. Stuyfzand, and H. Timmer. 1998. Water quality modeling at the Langerak deep-well recharge site, in *Artificial Recharge of Groundwater* J.H. Peters, et al., eds., Rotterdam, Netherlands: A.A. Balkema.

de Ruiter, Hannie, and Peter S. Stuyfzand. 1998. An experiment on well recharge of oxic water into an anoxic aquifer. *Artificial Recharge of Groundwater, Proceedings of the Third International Symposium on Artificial Recharge.* September 1998. Amsterdam: Balkema Press.

Fram, Miranda S., Brian A. Bergamaschi, Kelly D. Goodwin, Roger Fujii, and Jordan F. Clark. 2003. Processes affecting the trihalomethane concentrations associated with the third injection, storage and recovery test at Lancaster, Antelope Valley,

California, March 1998 through April 1999. United States Geological Survey Water Resources Investigations Report 03-4062.

Gao, J. 2010. "Data collection and statistical analysis of deep injection wells in Florida." Thesis, Florida Atlantic University.

John, David E., Joan B. Rose, and Amy Kamarainen. 2004. Survival of fecal indicator bacteria, bacteriophage and protozoa in Florida's surface and ground waters. Final Report of Microorganisms in Aquifers Study. West Palm Beach, FL: South Florida Water Management District.

Katzer, T., and K. Brothers. 1989. Artificial recharge in Las Vegas valley, Clark County, Nevada. *Ground Water* 27(1): 50–56.

Landmeyer, J.E., P.M. Bradley, and J.M. Thomas. 2000. Biodegradation of disinfection by-products as a removal process during aquifer storage and recovery. *Journal of the American Water Resources Association* 26:861–867.

Lucas, M., and K. McGill. 1997. Aquifer treatment methods for preventing iron concentrations in the recovered water from aquifer storage and recovery wells installed in clayey sand aquifers of the Atlantic Coastal Plain. Proceedings of AWRA Symposium, Conjunctive Use of Water Resources: Aquifer Storage and Recovery. American Water Resources Association. October 19–23, 1997, Long Beach, California.

Merrit, M.L. 1985. "Subsurface storage of freshwater in South Florida: A digital model analysis of recoverability." US Geological Survey Water-Supply Paper 2261.

Miller, C. T. et al. 1993. Fate of organochlorine compounds during aquifer storage and recovery: The Las Vegas experience. *Groundwater,* 31(3):410.

Mirecki, J.E., B.G Campbell, K.J. Conlon, and M.D. Petkewich. 1998. Solute changes during aquifer storage and recovery: Testing in a limestone/clastic aquifer. *Groundwater* 36(3):394.

Nellor, M.H., R.B. Baird, and J.R. Smith. 1984. Summary of Health Effects Study: Final Report. Whittier, CA: County Sanitation Districts of Los Angeles County.

Palmer, R., D. Wendell, A. Hutchinson, and T. Foreman. 1997. Optimization of ASR operations in a nitrate and PCE contaminated aquifer. Proceedings of AWRA Symposium, Conjunctive Use of Water Resources: Aquifer Storage and Recovery. American Water Resources Association. October 19–23, 1997, Long Beach, California.

Pavelic, Paul, Peter Dillon, and Brenton Nicholson. 2007. Formation and attenuation of DBPs in groundwater at ASR sites. *ISMAR6 Proceedings,* October 8 – November 2, 2007, Phoenix, Arizona.

Pitt, W.A.J., and S. Magenheimer. ASR technology: Avoidance and solutions to aquifer clogging problems. Proceedings of AWRA Symposium, Conjunctive Use of Water Resources: Aquifer Storage and Recovery. American Water Resources Association. October 19–23, 1997, Long Beach, California.

Pyne, D.G. 2004. Aquifer storage and recovery (ASR) issues and concepts. A position paper prepared by the St. Johns River Water Management District (SJRWMD). Jacksonville, FL: SJRWND.

Pyne, David G. 1995. *Groundwater Recharge and Wells: A Guide to Aquifer Storage Recovery.* Boca Raton, FL: Lewis Publishers.

Pyne, David G., P.C. Singer, and C.T. Miller. 1996. Aquifer Storage and Recovery of Treated Drinking Water. Awwa Research Foundation Report 90689. Denver, CO: Water Research Foundation.

Stuyfzand, Peter S. 1998. Quality changes upon injection into anoxic aquifers in the Netherlands: Evaluation of 11 experiments. Proceedings of the Third International Symposium on Artificial Recharge. Amsterdam: Balkema Press.

Singer, P.C., Mallikarjun Avs, Cass T. Miller, and Carolyn Mojonnier. 1993. Examining the impact of aquifer storage and recovery on DBPs. *Journal - American Water Works Association*, 85(11):85–94.

Texas State Water Board (TSWB). 2012. State Water Plan, 2012, Chapter 7, Water Management Strategies. www.twdb.state.tx.us/publications/state_water_plan/2012/07.pdf

Thomas, J.M., W.A. McKay, E. Cole, J.E. Landmeyer, and P.M. Bradley. 2000. The fate of haloacetic acids and trihalomethanes in an aquifer storage and recovery program. *Groundwater*, 38(4): 605–614.

USEPA. 1999. *Class V Underground Injection Control Study, Volume 21: Aquifer Recharge and Aquifer Storage and Recovery Wells.* Office of Ground Water and Drinking Water, EPA/816-R-99-014u.

Wendell, Daniel. 2010. Establishing reasonable expectations for the long-term performance of ASR wells. Proceedings of the AWWA California/Nevada Conference. October 2010, Sacramento, California.

This page intentionally blank.

Chapter 5

Planning and Construction of ASR Systems

Based on the lessons learned from the 204 ASR programs outlined in chapters 3 and 4, this chapter provides proven solutions for achieving injection, recovery, and water quality goals (success) with new projects, with the caveat that geological issues beyond the control of the engineers and hydrogeologists are the overarching factors in defining success with any ASR project. More importantly, instead of a design guide covering ASR well construction, this chapter provides basic guidelines for planning and implementing ASR systems, beginning with an overview of the concept, followed by a detailed discussion on how practitioners may approach a feasibility study for ASR systems, including an outline for implementation and guidance on how to continue the ASR feasibility phase to cover the pilot testing program and well construction. The chapter concludes with guidance on storage zone development, which leads to chapter 6, which addresses operation. This chapter discusses:

- Plan of study
- Feasibility study
- Pilot-testing program
- Well design
- Water quality
- Well construction
- Wellhead
- Well acceptance and pumping test
- Storage zone development

Each ASR system is unique. Specific design information will vary based on local, state, and regional regulations, as well as the existing geologic conditions and the water quality in the proposed aquifer zone and injection source. For details on well design, refer to AWWA Manual of Water Supply Practice M21, *Groundwater*; AWWA's publication *Siting, Drilling, and Construction of Water Wells;* and the AWWA Standard A100, Water Wells.

PLAN OF STUDY

As presented in chapter 1, ASR appears simple at face value. The basic ASR concept includes pumping water into a suitable underground formation and recovering it at a later time when needed. The concept sounds simple, but the process is not. First and foremost, a feasibility study is needed to consider factors such as water quality, underground conditions, local regulations, and economics. Candidate injection water quality sources must be carefully evaluated, as must the expected recovered water quality, with respect to intended use. Quantities and rates of injection and recovery must also be evaluated to assess whether or not ASR may be implemented economically.

Typical questions that should be addressed during the Plan of Study (POS) phase include:

- Is ASR legally allowed?
- What are the ASR program goals?
- What is the ASR program timeline?
- How much water is needed? For what purpose? Timing?
- What are the water sources for injection?
- How much treatment is required for feasibility?
- What funds will be used to pay for the program?

Experience has shown that allocating three to five years to complete the process provides the most informed decision making. The quick one- or two-year study is insufficient to fully understand the operational issues that may arise. The POS phase should be used to formally develop realistic time frames for implementation and data collection. Many questions will arise during design and testing, including laying out a format for how questions will be answered. A first step to address many of the questions about the potential for ASR is the development of a detailed POS.

A POS allows for the early evaluation of ASR. Realistic expectations are critical. The POS lays out the basic design of the ASR project, establishes metrics to assess and understand the usefulness of the project, and identifies the baseline performance expectations and what metrics will be used to measure that performance. For example, if an owner expects to achieve 100 percent recovery, how will recovery efficiency be calculated? Another example is understanding the time and amount of water required to fully develop or flush a storage zone, which is typically underestimated. Finally, designers and owners need a full understanding of pretreatment and posttreatment requirements.

As with any project, proper planning increases the chances of success. Owners are highly encouraged to invest adequate time, money, and resources into the planning process to assess the factors needed to deliver success. Issues on how regulatory and technical questions will be addressed, flexibility during testing, and the need for pilot testing should all be discussed in the planning phase of an ASR project. In addition, identifying questions and issues to address during the construction and testing phases helps provide direction.

Water withdrawal permits for testing and storage zone development may need to be secured prior to design. Pre- and post-treatment should also be addressed depending on the expected water quality of the injection and recovered water, mineralogy of proposed storage zone, and use of recovered waters. In many instances, injected water may adversely impact the composition of stored/recovered water (e.g., release of arsenic due to dissolved oxygen levels of injection water) on a temporary or permanent basis. Use of an existing supply well that could be converted and used for preliminary testing may accelerate the schedule, but caution should be taken as the design and construction of the existing well may not be suitable because supply wells are typically designed for long-term withdrawal of water, not injection.

FEASIBILITY STUDY

After the basic outline of the ASR project is defined in the POS, the Feasibility Study can begin. Proper planning during the Feasibility Study will likely result in cost savings and contribute to avoiding problems and delays during implementation. The Feasibility Study is generally the first step in the design process and provides early go/no-go decision-making information to owners. The Feasibility Study expands on the POS to address how data and questions will be addressed during construction and testing. Typically, a Feasibility Study provides a detailed plan of action for investigating underground conditions, collects data needed for final design, details an implementation strategy, and estimates planning-level costs. If additional data are needed, the Feasibility Study may recommend that a pilot-test well be constructed to provide site-specific information.

An important aspect of any ASR project is recognizing that ASR systems are not intended to produce *new* water, but rather to provide a cost-effective means to store and recover *existing* water. Depending on specific project goals, some degree of blending with native water may be acceptable and should be considered when addressing the performance of the ASR system.

Typical questions to address during the Feasibility Study phase include:

- Is ASR legally allowed?
- Is there a need, and what is it (define and quantify)?
  - Peak shaving (reducing peak demands for water on water treatment plants by supplementing with ASR recovered water)
  - System expansion deferment
  - Emergency supplies
  - Maintaining distribution system pressures during high-demand periods
- Is there a supply deficit during a portion of the year?
- How much water is needed during the deficit period?
- Where is the water needed?
- What is the quality of the water needed?
- What are the expectations (owner, designer, regulator, etc.)?

Preliminary Investigation ASR Tasks

- A background investigation to locate and identify construction details and hydrogeologic information from wells in the area

- A detailed literature search to locate reports and publications in the area of interest. Most areas may never have previously been investigated as zones suitable for ASR.
- A review of regulations governing the construction of ASR wells and the requirements for injection of potable and/or raw groundwater or raw surface water, reclaimed water, or some other source

Intermediate ASR Questions (once initial ASR questions are answered)

- Is there water available to implement ASR?
- When is the water available?
- How much water is there?
- Is the amount of water sufficient to meet the demand deficit?
- What is the quality of the excess water available for ASR use?
- Can the excess water be captured, stored, and recovered from an ASR well?

Detailed ASR Questions (once intermediate ASR questions are answered)

- What constitutes a suitable underground environment?
- Does a suitable underground environment exist?
- What is the storage zone mineralogy?
- Will there be geochemical reactions (native water quality and/or geology)?
- Can water be injected at a reasonable rate (rate and water quality)?
- Will injected waters impact adjacent users?
- Can the stored water be recovered (rate and water quality)?
- Is treatment needed prior to injection?
- Is treatment needed after recovery?

Planning Stage ASR Questions

- Where is the project located, and what are the characteristics of potential storage zone(s)?
- What are the applicable local, state, and federal regulations governing ASR systems in the project location?
- Is there overlying confinement?
- What is the transmissivity of storage zone?
- Is there underlying confinement?
- What is the areal extent of storage zone?
- What is the water quality of storage zone?
- What is the mineralogy of storage zone?

A general outline for an ASR Feasibility Study report follows as a tool for consideration in the development of a conceptual plan. The answers to the previous questions will provide the context, and the proposed use of the recovered water will define the need for post-treatment.

The following is a suggested outline for an ASR Feasibility Study report:

1. Introduction to the utility, environment, and needs for managing water
2. Project objective(s) (estimated injection, storage, and recovery quantities, rates, and quality)
3. Water demands (e.g., seasonal, annual, etc.)
4. Regulations
5. Literature search
6. Water supply quantity, quality, and availability
7. Potential recovery water quantity and quality
8. Storage zone characteristics
9. Alternatives analyses
10. Site selection
11. Construction and testing plan
12. Cost estimate
13. Implementation schedule

PILOT-TESTING PROGRAM

The pilot-testing program lays the foundation for the final design of the project. Formulation of a pilot-testing program begins with a clear understanding of what information needs to be collected. Data collection is crucial to obtain good information for decision-making. Numerous data points will be required, and collection of some data may not be achievable and/or economical with construction of a single well. For example, it would be more advantageous to collect core samples on a second well once the lithology is confirmed from the initial well. That way samples will be collected from the most likely ASR zones instead of at specific intervals, many of which may not be suitable. Information is needed for well construction, for assessment of confinement, to evaluate the hydraulic properties of the desired storage horizon, to answer regulatory questions, and for identification of water quality. (For a full detail of groundwater flow and calculations, drilling methods, well construction, and testing protocols, refer to M21; Standard A100; and *Siting, Drilling, and Construction of Water Wells*.)

The next step is to develop a plan that will allow collection of the information needed. A pilot well-testing program is developed by clearly defining and assigning the parameters needed to complete the test. The site access, staging areas, management of fluids produced during the drilling and testing process, and proper disposal of cuttings must be described so contractors understand the conditions for construction and testing, including any regulatory requirements. Possible space and noise constraints must be considered. The utility must investigate contractor licenses to ensure compliance and finalize any permits, temporary or permanent, prior to bidding, and address in the contract how the designer and contractor will issue reports on construction status to regulators and the owner. Items such as points of connection to potable water, cross-connection protections, sanitary facilities, and the impact of the program on water treatment plants, distribution systems, and pump stations need to be defined. The use and disposal of test water, as well as the location, quality, and amount of water required during construction and testing need to be confirmed prior to bidding. If further research of available records and literature is warranted, this phase allows additional time for such research.

Another consideration during the pilot testing and design phase is that of contactor qualification. Specialized testing is often required for data collection, so qualified contractors and equipment are generally required. Contractors should be prequalified based on prior experience in successful drilling of wells similar to those intended in the ASR program. The designer may establish minimum requirements for the contractor and equipment, and these requirements should be discussed with the owner's purchasing staff to confirm compliance with applicable regulations or ordinances.

Phased construction is generally preferred to maximize data collection. This involves drilling an initial bore, called a pilot hole, to gather preliminary hydrogeologic data, in advance of reaming and setting casing. Pump tests and geological logging of the formation should occur after the casing is set. Drilling one well, determining its operational parameters, and determining the geology should be performed prior to drilling additional wells. A second well, which may be used for the full-stage ASR, can provide added hydrogeologic data. The wells should be subjected to a series of injection/withdrawal cycles during testing to determine the likely parameters that need to be addressed as a part of permitting and operations. If possible, as the project progresses, the initial well should be converted into a monitoring well or an ASR well if it has a wide enough diameter.

Cuttings is the term used for the material from the native formation that rises to the top during the drilling process. Cuttings should be collected during drilling to determine the geology of the aquifer formation. The subsurface data collection process also includes geophysical logging of formation material, porosity, water quality, and other information. Specialty firms can be hired to help hydrogeologists and engineers understand the subsurface findings through a variety of methods that interpret the water and geology, most of which were developed for the oil industry. Refer to Bloetscher et al. (2007) for more details on geophysical logging.

In the borehole, a video survey should be conducted to allow the hydrogeologist, engineer, and staff to view the formation and get an idea of channels and vugs (voids) in the formation that might indicate water movement. This information is used to identify aquifer characteristics, such as hydraulic conductivity, porosity formation type, secondary porosity (interconnectedness of voids), and formation thickness, as well as water quality (both native and proposed injectate), including the modified fouling index (MFI), silt density index, recovered silts and sands, potential oxygen reduction potential and the like. Packer tests are usually required by regulatory agencies to verify well construction veracity, and pump tests.

WELL DESIGN

Figure 5-1 shows the typical parts of the well. The casing, well depth, and injection horizon are all important pieces of the system. The engineer will document the casing material and diameter. The equipment at the top of the well, or wellhead, includes gate and check valves and pressure recorders to monitor operations to ensure reliability of the ASR well. Note that none of these elements are significantly different than those found in standard water-supply wells, although some differences may be found based on local practice, regulations, and need.

Casings are pipes that prevent formation collapse. Based on the ASR survey outlined in chapter 3, the typical casing material is steel, although PVC and fiberglass are also used (Figure 5-2). The diameter of ASR wells varies from 5 in. to 42 in., but the casing diameters were typically 12 in. or 16 in. (Figure 5-3). Most steel wells are screened with stainless steel, and in California, cut casings are used (i.e., a tool is used to actually cut the casing pipe).

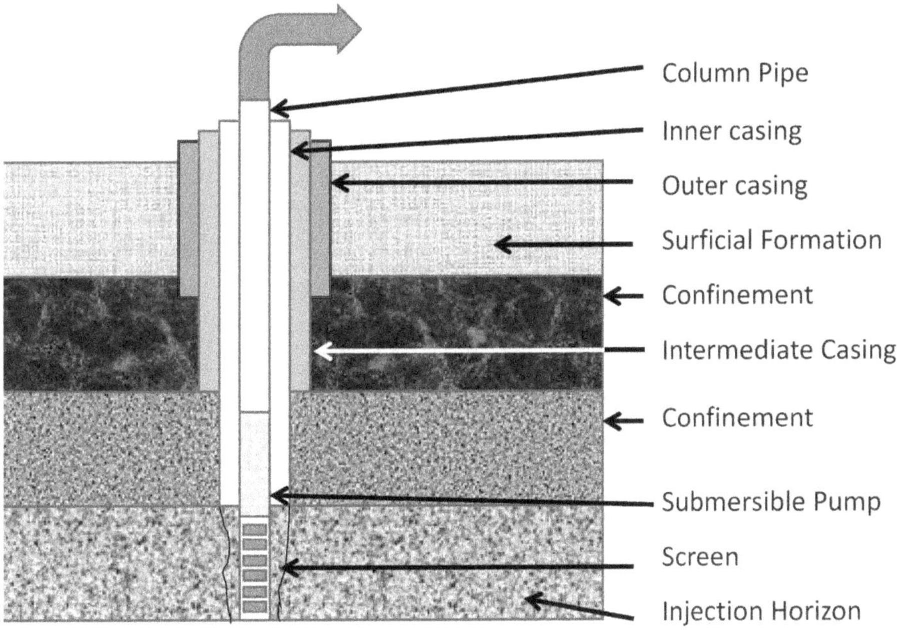

Figure 5-1 Parts of the ASR well; assume cementing between all casings for most ASR wells

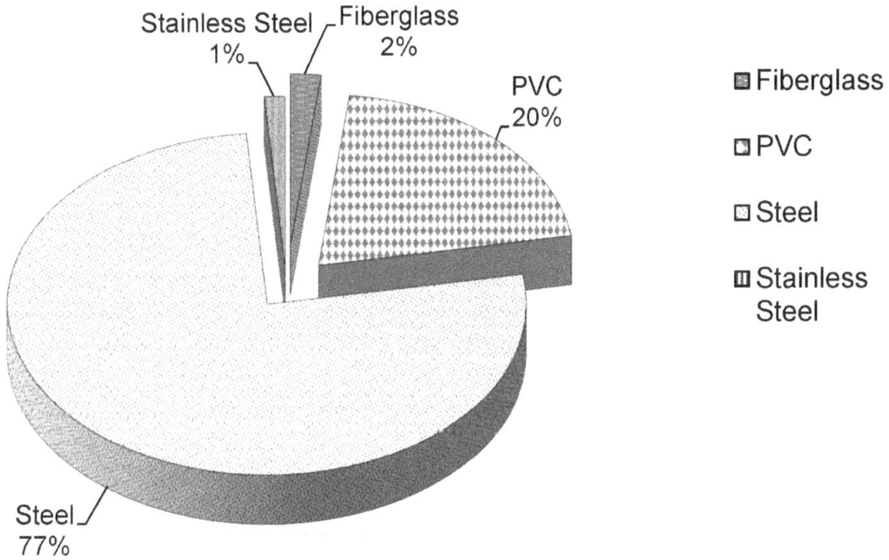

Figure 5-2 Casing material used in ASR wells

The major exceptions are the limestone wells in Florida, which are primarily open-hole (no screens). Several states have regulatory requirements regarding permissible casing settings for ASR wells.

Selection of the proper storage zone is key for a successful project. The drilling and testing phase should ensure that information needed for the designer is obtained, including overlying confinement parameters, the hydraulic conductivity of the targeted storage zone, and the presence of underlying confinement. This information will allow the designer to determine the optimal casing setting depth as well as the well depth.

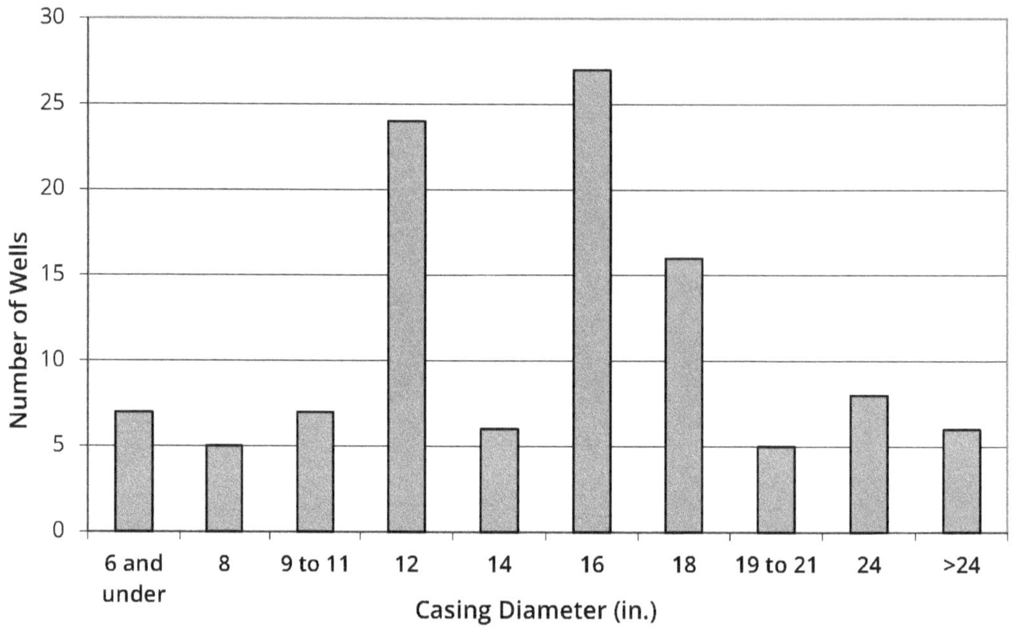

Figure 5-3 Casing diameter of ASR wells

The depths of ASR wells in the United States range from 90 ft to over 3,000 ft. Typical depths are less than 1,000 ft, with 400–600 ft the most common. Twelve percent of the surveyed wells are less than 200 ft deep, while 5 percent are over 2,000 ft deep. Almost all of the injection horizons are less than 1,000 ft thick, although the California well data does not lend itself to a clear understanding of the injection and recovery horizons because of the method used to cut the casing. By dividing the thickness of the injection horizons into increments of 100 ft, the majority of injection horizons were found to be less than 100 ft thick, which concurs with prior studies and literature (Bloetscher et al. 2005).

An assessment of the vertical permeability can be obtained from cores and packer testing. Laboratory analysis is useful to assess the vertical conductivity of the confining unit and verify overlying confinement. With respect to the storage zone, the horizontal conductivity is important and can be used in conjunction with data collected during withdrawal tests and packer tests to estimate the specific capacity of the storage horizon. As with the vertical conductivity, horizontal conductivity is measured in the lab, and/or by aquifer performance testing (see M21 for information on conducting aquifer performance tests).

Setting the final casing in a competent formation will optimize cementing and minimize the potential for channeling of water within the formation or up the borehole along the casing. In addition, the casing should be set as close to the base of the confining unit as possible. This will facilitate recovery efficiency as stored water will tend to remain within the storage zone and not migrate upward. Final casings set too deeply have had difficulty recovering stored water as injected waters have migrated vertically. Conversely, the final casing must not be set too shallowly, as this allows for a larger injection zone than needed, which could adversely impact recovery efficiency by dispersing the injected water too much. This is especially true when storing water in brackish to saline waters because the freshwater will float to the top of the aquifer. Because recovery pulls from the entire injection horizon, brackish water is pulled into the withdrawn freshwater in such installations.

Next to safety, the most important component of the drilling and testing program is flexibility. Since little is initially known about the site-specific underground conditions, flexibility in the drilling and testing program requirements should be understood by all parties (i.e., owner, contractor, engineer, and regulators). To accomplish this, the contract wording should be carefully structured to allow changes to accommodate unforeseen conditions encountered during construction.

The data requirements to include in the construction specifications or to be defined by the hydrogeologist during construction are as follows:

- Well diameter
- Proposed total depth
- Position of the screens with respect to a fixed reference point (e.g., top of flange)
- Portion of the open hole if constructed in rock
- Method of construction and materials to be used for the screen and casing
- Pump design (i.e., withdrawal capacity and head)

WATER QUALITY

As noted at the end of chapter 4, the use of ASR technology in brackish aquifer systems presents challenges related to the mixing of the injected freshwater with the native brackish water. However, according to the ASR survey, the total dissolved solids of the original water are often under 500 mg/L, which eliminates the density differential issues associated with vertical migration of the freshwater that reduces recovery in some of the south Florida wells that inject into brackish aquifers (Figure 5-4). This means that for many of the formations, freshwater is being injected into a freshwater aquifer, so buoyancy is not a factor and typically results in the potential for high recovery.

This does not mean water quality is not an issue, however. As noted in chapter 4, metals, carbon dioxide, and DBPs are potential water quality concerns that might impact any project. The mixing of native and injected waters with different concentrations of dissolved oxygen and temperatures, as well as oxidation/reduction chemistry issues between the native and injected water, can lead to water quality changes in the aquifer (see chapter 4). ASTM standard D-4195-08, Standard Guide for Water Analysis for Reverse Osmosis and Nanofiltration Application, requires water analyses to be performed on raw water for the following selected parameters (Bloetscher and Witt 2000, Bloetscher et al. 2005):

- Calcium (Ca)
- Magnesium (Mg)
- Sodium (Na)
- Strontium (Sr)
- Barium (Ba)
- Aluminum (Al) (total and dissolved)
- Manganese (Mn) (total and dissolved)
- Iron (Fe) (totaled, dissolved, and ferrous)
- Potassium (K)
- Bicarbonate (HCO_3^-)
- Sulfate (SO_4^-)
- Chloride (Cl^-)
- Carbonate (CO_3^-)
- Nitrate (NO_3^-)

- Fluoride (F^-)
- Phosphate (PO_4^-) (total)
- Silica as silica dioxide (SiO_2^-) (total and dissolved)
- Total dissolved solids (TDS)
- Total organic carbon (TOC)
- Hydrogen sulfide (H_2S)
- Free chlorine (Cl_2)
- Oxygen (O_2)
- Carbon dioxide (CO_2)
- pH
- Temperature
- Turbidity (nephelometric method)
- Silt density index

Temperature, pH, and the silt density index should be measured on site when the sample is collected, and the depth from which the water was retrieved should be noted. The water should also be tested for ammonia.

The sampling period should be determined by the stability of the parameters in the well, i.e., how often, if ever, the parameters change in concentration and the amount of the change. Changes in water quality are an indicator of problems with wells and, if the changes are large enough, further treatment of the recovered water may be required. In addition, certain changes in water quality can cause adverse environmental impacts, such as metals recovery.

While analyses for many parameters should be performed by a qualified chemist in a state-certified laboratory, field probes can be used to obtain reasonably accurate readings for routine testing (pH, temperature, TDS, conductivity). Testing for other chemical and physical parameters may be required in certain regions depending on specific groundwater quality of the area. Given the potential for multiple water quality issues, a professional familiar with geochemical issues and who understands state and federal guidelines on any ASR project should be engaged.

WELL CONSTRUCTION

Construction and testing of ASR wells is different than that of production wells and is a labor-intensive task. As discussed previously, a drilling contractor for the construction and testing of the system should be selected based on qualifications and price. Prior to contractor mobilization, preconstruction photographs and videos should be taken of the area to be drilled and existing facilities. A preconstruction meeting with all parties should be held to discuss project goals, requirements related to the construction sequence, and the proposed testing plan.

One item that is often overlooked is the establishment of vertical and horizontal controls. Horizontal control refers to a coordinate system (latitude–longitude, state plane coordinates, etc.; a surveyor can help with this task). Vertical control requires an elevation benchmark. The owner should establish a temporary benchmark that defines both vertical and horizontal control at the project site. Establishing this control is crucial for accurate measurement of water levels and location of permanent facilities.

Construction materials are important, especially in environments with brackish or saline waters. The contractor and engineers need to review shop drawings for equipment

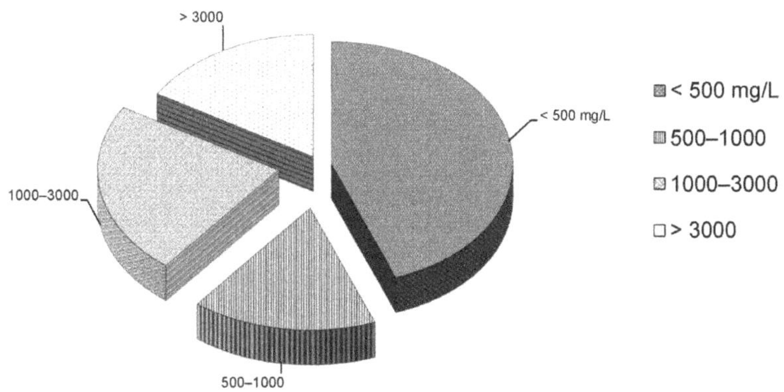

Figure 5-4 TDS of the injection horizon for Florida wells that inject into brackish water

and materials carefully to ensure compliance with the contract documents; review flow measurements and gauges for data collection and install in accordance with manufacturer recommendations; and complete shop drawings and manufacturer cut sheets, and confirm in the field upon installation. In many instances, the contractor and manufacturer are required to submit a certification of proper installation and materials. Such certification could be of value once the project is completed to facilitate the resolution of any issues with equipment or materials.

Both the owner's representative and the contractor should monitor construction and record testing data daily; such activity may also be required by regulatory agencies. The contract documents should require the contractor to submit daily notes to the owner's representative on a weekly, if not daily basis. For payment purposes, a schedule should outline the quantities and costs of each component in the construction process (e.g., drilling, casing, cement), as well as who is responsible for obtaining and paying for these items. These materials should be referenced when used in the daily records of construction and testing to avoid discrepancies during review of payment invoices.

Collected data in electronic and hard format needs to be submitted to applicable parties to facilitate review, with deadlines on how long the designer/owner/regulator have to review and evaluate data prior to decision making (e.g., setting casing depth), and specific timelines in the sequence and constraints section of the contract documents need to clearly identify the potential for standby time to the contractor. Contract documents provisions must assign responsibility to the contractor for identifying potential schedule impacts, allowing the owner to adjust the schedule to avoid being responsible for standby costs.

WELLHEAD

An ASR wellhead can be similar to the wellhead for a regular well, except that it must have the appropriate valving and pressure sensors to allow for injection and withdrawal. Figures 5-5 to 5-7 illustrate different types of ASR wellheads. Figure 5-5 is a simple system in which submersible pumps and valving allow injection and withdrawal. Check valves and actuated gate valves can be used to direct flows. Figure 5-6 shows a vertical turbine pump on an ASR test well. The vertical turbine can be used for injection and withdrawal. Again, proper valving and piping are important for flow. Manual changes to the configuration can be used to ensure the system operates correctly. Figure 5-7 shows a system where an unused water supply well acts as an ASR well. At this site dual piping and valving is used to inject and withdraw water. Determining the amount of automation desired by the client is an important consideration in designing the wellhead. Figure 5-8 shows a typical ASR wellhead design used by many systems. Local requirements may vary.

Figure 5-5 Basic configuration for injection piping and withdrawal piping for submersible pump

Figure 5-6 Basic configuration for vertical turbine pump (in test mode)

Figure 5-7 Basic configuration for injection piping and withdrawal piping for vertical turbine pump

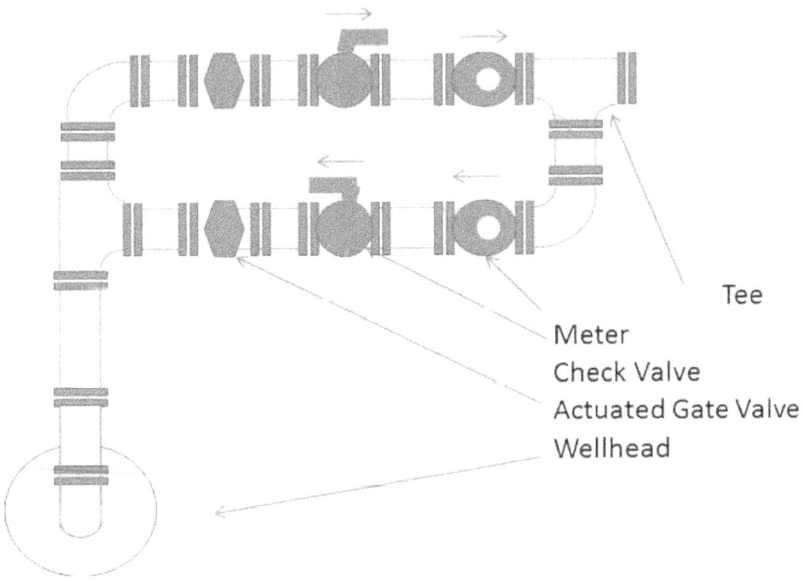

Figure 5-8 Components of a wellhead

WELL ACCEPTANCE AND PUMPING TEST

A newly constructed well and pump should be evaluated as a unit to establish standard performance. Data should be collected to determine water losses that relate to well design and construction. Comparative data pertaining to the physical condition of the pump unit should also be collected. Proper study and comparison of such data enable the operator (or consultant) to anticipate maintenance and repair needs. After any repair or maintenance work, similar tests should be repeated. This data should include the following:

- Water level and water quality measurements made before, during, and after the (drawdown) pumping test
- A record of the pumping rate
- Hydrographs generated during the test
- Any other field data collected (manual or computer generated)
- A copy of the hydrogeologist's report on the procedures
- A copy of all results from core, cutting, water quality, pump tests, etc.

STORAGE ZONE DEVELOPMENT

The startup procedures used for most ASR systems replicate the proposed full-scale test on a shortened cycle. The initial pumping may be reduced as well. However, development allows for equipment evaluation, monitoring, and debugging, especially control electronics. Development may be the most overlooked issue in well construction, but good development practices will provide an efficient well with long life and low maintenance costs. Proper development of the storage zone entails pumping the well until the water runs clear of silt, sand, and other underground matter and the formation is no longer releasing material. The design and construction personnel should agree on how the storage zone will be developed and under what conditions full development is achieved, based on the specific project requirements.

Prior to commencement of development, establish metrics to monitor progress and facilitate well performance. Basic production well strategies should include measurement of specific capacity and water quality parameters such as TDS, sand content, and silt density index. For new well sites, the exact conditions of the potential storage zone may not be well known, so flexibility may be needed to adjust well development strategies based on the conditions encountered. It is not unreasonable for a contractor to take 30 days to fully develop the storage zone prior to cycle testing because clogging is a common issue with ASR wells. The same problems that occur with water supply wells will occur with ASR wells, only more so because of the movement of fines within the formation caused by the cycling of water in and out. Wells that are not developed adequately or properly can lead to program failure.

Downtime can be minimized by following basic procedures in advance of full-scale testing. Conduct a preliminary system evaluation prior to final completion, where the contractor demonstrates that all systems are operable, either manually or automatically, prior to testing. This preliminary inspection should also include monitoring equipment. A related check should include confirmation of procedures with the owner and regulators, if applicable, to ensure full compliance prior to full-scale testing.

During well development, the designer or engineer should monitor specific parameters to confirm proper operation of systems. These parameters will vary depending on regulatory input, design recommendations, geology, and local issues. Common items to track are flow (in/out), silt density index (SDI) for turbidity and suspended solids, and

TDS or conductivity (saltwater). A site-specific checklist should be developed to track that items are completed according to contract documents and permits. Communication with appropriate parties, including but not limited to the owner and contractor, is essential. The engineer should establish a protocol for managing data to facilitate review and reporting. Spreadsheets and graphs are easy tools for reporting and evaluation of data like flow rates, conductivity, SDI, time, temperature, and other parameters. The overall process is as follows:

Basic pre-startup checkout procedures:

1. Verify pumps are operating correctly
2. Verify check valves are operating correctly
3. Verify actuators are operating correctly
4. Verify sensors are operating correctly
5. Verify flow meters are operating correctly
6. Verify monitoring systems (computers) are operating correctly
7. Verify monitoring well systems (if applicable) are operating correctly

During development, flowmeters and TDS/SDI or other sensors must be operating and recording information correctly, that water quality (as applicable) is being monitored and any changes in quality receive an appropriate response, and that an active data management system is in place to capture, store, and disseminate all relevant information.

Not everything will go smoothly, so some modifications may be required based on the information collected during preliminary testing and startup. If this happens, the designer should have a plan to address "what if" scenarios with appropriate actions. Modifications to testing plans should be shared with owners and regulators as applicable.

Once the well is developed, recharge rates and recovery rates can be evaluated further. Sometimes recharge pressures are higher than anticipated due to chemical reactions in the storage zone or movement of fines. The same is true for recovery as stored water may have reacted with native water and altered the composition. Flushing of an entire storage zone usually takes months and sometimes years, so patience is required and expectations must be managed. Ideally, recovered water can be blended and put to beneficial use during the testing phase. Many brackish water wells will have lower recovery rates than injection rates. The reverse may be true for systems in aquifers that have been drained previously.

Unlike potable water supply wells, ASR wells must undergo a series of injection and recovery cycles. It is during these storage and recovery cycles that information is discovered about water quality issues, clogging potential, and geochemistry.

In the review of 204 ASR wells in the United States, operational systems tend to have had more than 10 cycles of injection/withdrawal, while test sites or inactive sites tend to have had less than five cycles, which means that the feasibility of operation was determined relatively early in the process (Figure 5-9).

Operationally, the well injection rates (in mgd) are predominantly less than 1.5 mgd. The most common range is 1.0–1.5 mgd, followed by rates just under 1 mgd. Very few systems inject, or attempt to inject, more than 2 mgd. An attempt at injecting 5 mgd was made in several south Florida wells, but was ultimately unsuccessful. Withdrawal rates tend to be slightly higher than injection rates, but the ratio of injection to withdrawal is approximately 1:1. With actual withdrawal rates, the most common volume of withdrawal rates is 1.0 to 1.5 mgd, followed by just under 1 mgd. South Carolina's wells are good examples where recovery is greater than injection in all cases. In brackish aquifers, the tendency is just the opposite—recovery is less than injection. Total recovery for a given site, typically

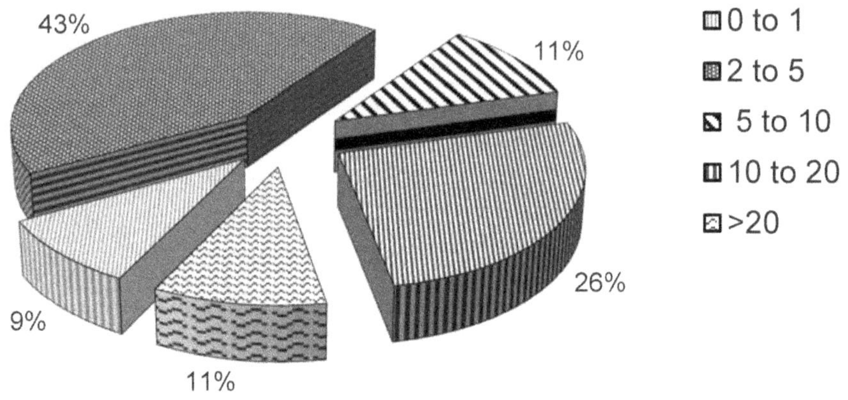

Figure 5-9 Number of cycles of injection/recovery per site

only one or two wells, was between 1.5 and 2 mgd. Looking at the amount of water stored, most systems store around a billion gallons, but can range to 10 billion gallons for larger, older systems.

DESIGN AND CONSTRUCTION RECORDS

The following should be obtained after the drilling process is complete and the casing and well screens have been put in place:

- Detailed individual well (geologic) logs
- Pump-design selection, discharge pressures, and setting
- Water-quality results for native groundwater and injected and recovered water
- Static (non-pumping) water levels in the aquifer, and if applicable, for overlying and/or underlying aquifers

When the well has been constructed, record drawings of the well should be made. These records should include:

- Casing, screen and annulus seal diameters, materials and settings
- Injection piping settings and diameters, and valving
- Total drilled and completion depths
- Screen setting(s) (or open borehole interval if constructed in bedrock)
- Method of construction and materials used for the screen(s) and casing(s)
- The driller's log of the materials encountered during drilling
- Slot sizes of the screen(s)
- The depths (settings) of the casing(s) and screen(s)
- Surveyed location of completed well(s)
- Surveyed elevation of water level measurement point

Pump data should include:

- The type, make, and model number of the pump installed
- The type and horsepower of the motor (driver)

- The pump setting (depth of the pump intake)
- The setting of the air line or other device for measuring the water level in the well
- Notation for the point (and reference elevation) used for measurement of the water level
- All information provided by the pump and motor manufacturer, such as capacity and efficiency data to include the total dynamic head (TDH), diameter and number of bowls, pump speed (rpm), etc.

An operations and maintenance manual should be submitted to the utility with all of this information.

REFERENCES

AWWA (American Water Works Association). 2014. Manual of Practice M21, *Groundwater*, 3rd ed. Denver, CO: AWWA.

AWWA. 2006. Standard A100, Water Wells. Denver, CO: AWWA.

Bloetscher, Frederick, and Gerhardt M. Witt. 2000. Effects of silt, salt and other fouling mechanisms affecting groundwater treatability. Proceedings of the American Water Works Association Annual Conference and Exposition. June 2000, Denver, Colorado.

Bloetscher, Frederick, Albert Muniz, and Gerhardt M. Witt. 2005. *Groundwater Injection: Modeling, Risks, and Regulations.* New York: McGraw-Hill.

Bloetscher, F., A. Muniz, and J. Largey. 2007. *Siting, Drilling, and Construction of Water Wells.* Denver, CO: AWWA.

This page intentionally blank.

Chapter 6

Operation and Performance Monitoring of ASR Wells

Once an ASR well is designed and constructed, its performance (flow rates) and water quality must be routinely monitored to meet regulatory requirements and ensure optimal operating performance. Performance monitoring should, at a minimum, focus on the timeliness and effectiveness of routine maintenance related to the operation of pumps and motors, and include recordkeeping of injection and withdrawal/recovery volumes and related water-level data in the aquifer. Water quality monitoring and recordkeeping are generally straightforward and often dictated by regulatory agency requirements. Monitoring for compliance with owner goals is specific to each project and may be more detailed than the regulatory reporting requirements. However, ongoing issues with well inefficiency from biofouling and encrustation are common with most wells, so monitoring and recordkeeping of these conditions will be covered in more detail.

DATA MANAGEMENT

Adequately monitoring and identifying performance issues requires gathering and compiling comprehensive data on equipment and hydrogeology. These data help detect a loss of performance efficiency—evidenced by a loss of water production, reduced pump or motor efficiency, increased power demands, and excessive drawdowns—as well as the related cause(s) for reduced performance. Collected data are also used to develop an appropriate proactive maintenance schedule to avoid breakdowns and irreversible declines in well efficiency, identify the full cost of water production, and determine the need to schedule capital improvements. The data collected should include the parameters discussed in chapter 5.

Recordkeeping should conform to other records used at groundwater and surface water plants for chemical usage, pump run times, and so on. Added information to collect includes, but is not limited to, information on well water levels during recharge and recovery, pumping rates during pumping and recovery, water quality data, backflush frequency, specific capacity, and specific capacity trends for recharge and recovery.

Records must be collected in an orderly and reproducible manner, regardless of the form used. Record the date and time for each set of measurements. Other information to record includes design, construction, and operational data, which are discussed in detail in this chapter.

A key objective of recordkeeping is compiling information that allows comparison of current operating characteristics and conditions with the original calculated (theoretical) design performances. This data also helps in assessing the cause of problems with the well and pumping system and in developing design modifications to improve efficiencies and production capability in future wells. Injection wells require collecting more water quality data and performing specific capacity tests to gather information on clogging and geochemistry issues more frequently than required for wells that only withdraw water.

Trending

Pumping data from production wells must be collected to compile the operating history and to schedule well maintenance. This data can also help with evaluating the cost of water production and guiding improvements to the physical system and operating methods. Capacity declines occur in most water wells primarily because of gradual clogging within and immediately adjacent to the well screens. Tracking specific capacity helps identify clogging problems.

The total pumpage volume for each well should be recorded monthly, and graphed to illustrate the seasonal and yearly production rates. This data can be used for projections of water withdrawal rates, to monitor the actual volume of water produced from each well, and to predict allowable operating periods between maintenance shutdowns. This data can usually be recorded from a totalizer on flowmeters installed in the discharge piping of each well.

Record the water levels in the well during non-use (static) periods and pumping (dynamic) cycles, along with associated flows. Measuring water levels in the aquifer over time provides a baseline for determining the amount of drawdown and the amount of increased pressure (draw-up) associated with a given flow rate. This allows for calculation of the well's specific capacity, which is the ratio of the yield of each well to its drawdown, expressed in gallons per minute per foot of drawdown (gpm/ft). By plotting over time, the operational trend of each well will be apparent. When the specific capacity trends downward, repairs can be scheduled. Trends in the water level during pumping over time can reflect losses in efficiency in the well production.

Water level measurements in observation and monitoring wells can also be used to evaluate water levels in the aquifer at any given point. Monitoring pumping levels for any given well illustrates the seasonal variations of pumping rates, adjacent water body level, water temperatures (well water and adjacent water body), and drawdown with time. These measurements can help hydrogeologists and engineers determine the *hydraulic conductivity,* or the ease with which water flows through the aquifer. Hydraulic conductivity directly determines how productive a well might be. A more common term, *transmissivity*, is the hydraulic conductivity multiplied by the thickness of the production zone.

Specific capacity depends not only on the transmissivity of the aquifer but also on construction factors such as screen type, well diameter, degree of aquifer penetration, and degree of well development. In general, when a well is new or new screens have been installed, the specific capacity is expected to fluctuate with pumping rate, adjacent water body level, temperature, aquifer water levels, and degree of screen efficiency. Refer to Manual M21 (AWWA 2014) for more specifics on well testing, as all wells have the same specific capacity issues.

Water quality is difficult to define in a guidance manual such as this. However, a variety of parameters should be routinely gathered to identify chemical interactions that occur below the surface as a result of the non-native water chemistry's interaction with the water in the aquifer formations, which can change the chemical composition of both (geochemistry). The geochemistry should be tracked for several cycles and over several years to ensure no interactions are occurring that would contaminate the recovered water or require treatment that was not expected. Metals leaching (discussed in chapter 4) is an example of a geochemical issue that impacts recovery. Likewise, water quality in the aquifer should be monitored and tracked to confirm that no changes to the formation are occurring as a result of injection.

Pump and Motor Installation and Maintenance Considerations

Decisions for specifying pumps for installation and for scheduling maintenance are typically based on initial cost, installation cost, cost of operation, cost of maintenance, and expected equipment life. Properly selecting pumps and motors for wells initially can help system performance, but changing well-performance and/or hydrogeologic conditions sometimes justifies changing components to maintain efficient and economical operation.

An example of a typical performance curve for a vertical turbine pump, the type most commonly used for ASR wells, is shown in Figure 6-1. A curve such as this one is used to check the observed range of pump operating conditions against those provided by the manufacturer. The results of this comparison are used to ensure reliable operation when evaluated along with ambient air and water temperature ranges, pressures, flow, corrosive and abrasive water-quality indicators, power demand variation, duty cycle, and the functioning of protective devices. For optimum performance and life span, the pump should be installed according to manufacturer's instructions. Likewise, routine maintenance specified by the manufacturer, such as greasing bearings, polarity checks, and alignment checks, should be followed.

Continuous or long-term operation of a pump is generally preferable to intermittent or periodic operation. Most ASR wells pump for a period of days (or months) to inject water, followed by an idle period of days or months—and in some cases years—without additional pumping (in or out). The operation of an ASR well and the varying water demand usually requires some combination of off-and-on operational time and flow throttling. While long downtimes may not be too damaging, short on/off cycles may contribute to excess wear on equipment that is designed for consistent operation. To prolong pump life and well performance, system components and storage capacity should be designed to minimize the number of pump starts and stops per day. At the same time, the pump must be sized and set to preclude its operation for even a few minutes at "no delivery" to avoid damaging the pump bearings, which can lead to a motor overheating.

If the resulting well drawdown or the delivery system could cause the pump to run at little or no flow, protection should be provided to automatically cause pump operation to cease. Appropriate protection measures include installation of a level switch or

PUMP DATA SHEET Turbine 60 Hz

Company: Customer:
Name:
Date: 06/09/10 Order No:

Pump:
Size: 8FDLO (2 stages)
Type: Lineshaft Speed: 3450 rpm
Synch speed: 3600 rpm Dia: 5 in
Curve: E6208FGPC0
Specific Speeds: Ns: 3750
Pump Notes for Standard Sizes:
Suction Size-6" Discharge Sizes-5",6". Curves are certified for water at 60°F only. Consult factory for performance with any other fluid.
Vertical Turbine: Bowl size: 7.5 in
Max lateral: 0.56 in
Thrust K factor: 9.9 lb/ft

Pump Limits for Standard Construction:
Temperature: 120 °F Pressure: 400 psi g
Sphere size: 0.5 in

Search Criteria:
Flow: 719 US gpm Head: 150 ft

Fluid:
Water Temperature: 60 °F
Density: 62.25 lb/ft³ Vapor pressure: 0.2563 psi a
Viscosity: 1.105 cP Atm pressure: 14.7 psi a
NPSHa: — ft

Motor:
Standard: NEMA Size: 40 hp
Speed: 3600
Sizing criteria: Max Power on Design Curve

— Data Point —
Flow: 719 US gpm
Head: 150 ft
Eff: 81.3%
Power: 33.5 hp
NPSHr: 25.7 ft

– Design Curve –
Shutoff Head: 234 ft
Shutoff dP: 101 psi
Min Flow: — US gpm
BEP: 81.4% eff
@ 724 US gpm
NOL Pwr: 33.6 hp
@ 676 US gpm

– Max Curve –
Max Pwr: 34.5 hp
@ 681 US gpm

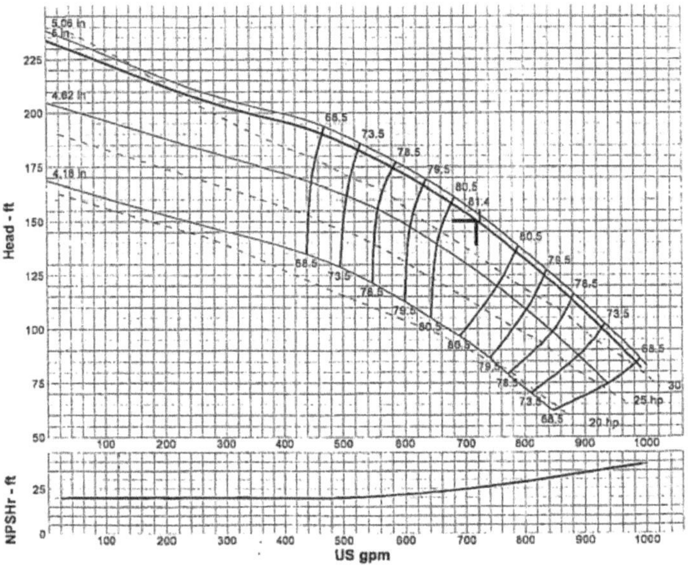

Performance Evaluation:

Flow US gpm	Speed rpm	Head ft	Efficiency %	Power hp	NPSHr ft
863	3450	117	78.8	32.4	31.9
719	3450	150	81.3	33.5	25.7
575	3450	176	77.9	32.7	21.4
431	3450	193	64.7	32.2	20
288	3450	204	48.1	30.6	20.2

Turbine V9 Selected from catalog: Goulds Lineshaft 60HZ Vers: 3.31

Figure 6-1 Pump design curve and data
Source: *Goulds Water Technology*

SCADA-system-connected transducer that shuts off the pump or sounds an alarm if the flow rate or pumping level drops below a predetermined safe minimum level.

Monthly Injection and Pumpage

The total and per-cycle injection and recovery volumes for an ASR well are generally required by permit to be regularly recorded and reported to the regulatory agency using a specific format (Table 6-1). Besides meeting regulatory requirements, the reported water quantities and corresponding seasonal and yearly production rates can be used to project future water withdrawal rates, provide comparisons between injection and recovery volumes, and predict maintenance intervals for the ASR well (AWWA 2014).

Table 6-1 Typical well and pump operational record form

WATER-LEVEL DATA							
Well	Date	Pumping Rate (*gpm*)	Static Depth to Water (*ft*)	Pumping Depth to Water (*ft*)	Drawdown (*ft*)	Specific Capacity (*gpm*/*ft*)	Pressure (*psi*)

Inline meters capable of instantaneous flow (e.g., gallons per minute) and total volume (e.g., gallons) measurements of in-and-out pumping of an ASR well should be located where viewing and calibration are easily possible. Separate, dedicated meters to discretely measure the inflow and the outflow of the respective delivery and recovery piping are appropriate. Utilities should avoid the use of a single meter capable of running in both directions (forward and backward) because the reverse flow cancels itself out and registers incorrectly on the meter. Meters that can communicate with a SCADA system are preferred.

Water Levels

Besides injection and recovery flow rates and volumes, operators should record water levels or the piezometric surface in the well during static and dynamic periods to determine the amount of corresponding mounding (increase in aquifer head) and drawdown during injection and withdrawal cycles, respectively. The static levels can identify changes in the amount of water that may be available in the aquifer under given conditions. Static and dynamic water-level data allows for the calculation of specific capacity and specific injectivity (i.e., gpm/ft of rise), which in turn can be used to monitor ASR well performance. The use of this data can be further enhanced when supplemented with similar water-level measurements collected in nearby observation and monitoring wells (Bloetscher et al. 2007).

Water Temperature

The temperature of injected, ambient (background), and pumped water should be recorded to establish baseline and operational trends. Variations in groundwater temperature can potentially cause the capacity of a well to vary as a result of changes to water viscosity.

Well Efficiency

The differential between the mounding or drawdown water level inside the well and in the immediately surrounding aquifer (typically <100 ft) as measured in nearby monitoring or observation wells provides a direct measurement of the operational efficiency of the ASR well (Figure 6-2). Therefore, besides the ASR well, at least one groundwater-level measurement location (commonly a monitoring well) is required. Measure and record the differentials from a designated location (e.g., inner casing rim) regularly.

Aquifer plugging and aquifer subsidence can affect specific capacity of the well and the resulting differential between the ASR well and the monitoring well(s). When a well is new or recently rehabilitated, the differential should be minimal because the well screen

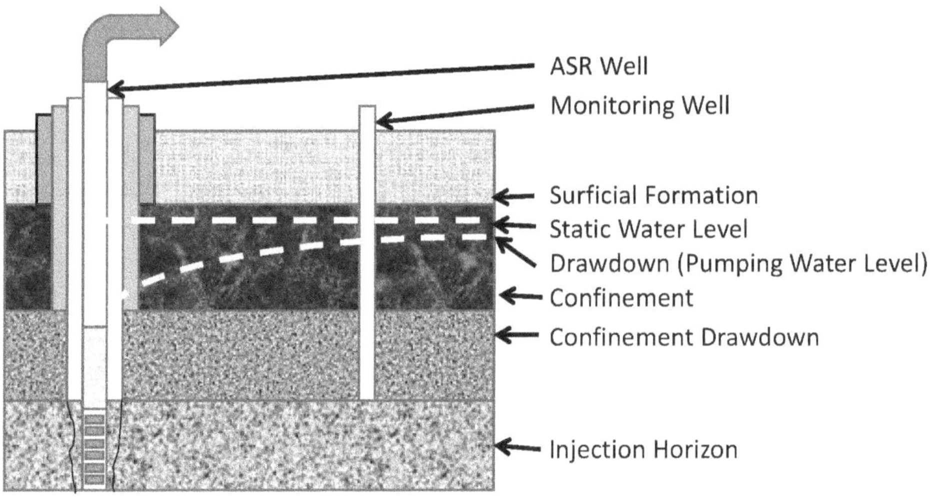

Figure 6-2 The components of pumping-induced drawdown at a well related to aquifer hydraulics and well design

and surrounding gravel-pack filter (if present) should be relatively free of obstructions. As the formation and/or well screen become plugged over time, the differential can be expected to increase as efficiency decreases from increased head losses in and around the well screen. A good starting rule of thumb is to implement rehabilitation measures when efficiency decreases by 15 percent when compared to the initial efficiency of the well.

Specific Capacity and Specific Injectivity

The specific capacity or the ratio of the pumping rate to drawdown is commonly used to establish and monitor the performance of a well. Specific capacity is calculated by dividing the pumping rate (gpm) by the observed drawdown (ft) at any specific time. To be useful as a measure of well performance, the well must be operated using approximately similar pumping rates and pumping durations. A bigger pump or deepening a well might give different results. Specific capacity is reflective not only of local aquifer hydraulics (e.g., the transmissivity of the aquifer) but also of intake design and construction factors, such as screen type and diameter, degree of aquifer penetration, and completeness of initial well development. Other factors that may influence specific capacity are pumping rate and duration, stage in hydraulically connected surface-water bodies, water temperature, aquifer water levels, and degree to which the screen impacts flow.

Similarly, specific injectivity (ratio of the injection rate accepted by the well to its water-level rise expressed in gpm/ft of rise), is also used to monitor the performance of an ASR well. Specific injectivity values are influenced by the same aquifer and well-construction factors that affect specific capacity. In addition, specific injectivity may vary with the corresponding injection cycle if clogging, geochemical reactions or other issues are occurring in the subsurface.

Specific capacity and specific injectivity values can help operations personnel evaluate catastrophic failures or long-term declines in well performance, particularly if formation or mechanical problems are the cause of capacity decline. Operators should calculate the specific capacity and specific injectivity of an ASR well quarterly, or at least annually, to identify maintenance needs in a timely manner.

Typical reasons for decline in ASR well performance, as reflected by decreases in specific capacity and specific injectivity, include changes in the formation, changes near the well bore (compaction, sand entrainment, subsidence) and well screen and/or gravel pack plugging caused by biofouling and/or chemical precipitation. Other ASR well performance problems may be related to:

- Changes in the hydraulics of the water-bearing zone (e.g., strata slumping—downward movement of the production zone when water is removed—fracture shifting/collapse)
- Incomplete development of the well at time of construction
- Pump wear and component breakage

Proper comparison and study of specific capacity and specific injectivity, along with groundwater-level differential measurements (if available), can establish trends for projecting maintenance and repair needs. Comparative data (e.g., pressure, power consumption) pertaining to the operation of the pump unit should also be collected. These data should include, but are not limited to:

- Water level measurements made before, during, and after pumping and injection tests, or system startup/design tests
- A record of pumping rates
- Hydrographs generated during testing

Operators should record data and assess the effectiveness of well maintenance activities to help project future maintenance needs, identify causes of well decline, and plan annual maintenance budgets. These records should include:

- The dates that maintenance was performed
- Triggering event(s) of reactive maintenance
- Results of pre- and post-maintenance pumping tests
- Methods (and materials) used in the maintenance procedures
- Other factors such as the color of the pumped water, amount of sand removed, odors, water quality analyses, downhole video inspection, and pump and piping condition

OTHER OPERATION AND PERFORMANCE CONSIDERATIONS

Pump suction breaks, excessive pumping of the formation (withdrawal or injection pumping), and increasing rates of drawdown can damage the pumps, piping, and aquifer formation. Silt and other materials can be dislodged. Operating pumps under conditions that afford a break in suction can cause damage to both the pump and the well intake zone because of cavitation and depressurization-induced turbulence.

Pump and injection surging water levels in an ASR well may affect casing and intake zone integrity if the well was not properly developed and stabilized during construction. Any surging action can stimulate sand, silt, and colloidal migration or dislodge corrosion products and precipitates. Pumping and injection can cause air entrainment in the pumps, which can damage the system piping and pump impellers. To this end, injection piping should be installed below water levels during withdrawal cycles (drawdown levels). Avoid closing valves to throttle flow rates. Instead, remove a pump bowl (stage) or slow the motor speed (e.g., use variable frequency drives, allowing the pumping rate to be increased or decreased safely). Lightning strikes and poor grounding may cause electrical surges that damage motors and pumps, so appropriate lightning attenuation should be installed where appropriate.

WELL ABANDONMENT

Regulatory agencies generally require utilities to abandon wells no longer in use. Wells must be abandoned in accordance with the regulatory guidelines. The following is a general outline of a well abandonment plan, subject to modification based on the nature and cause of the abandonment.

- Suppress the head of the water in the well by pumping a solution of sodium chloride and/or barium sulfide (barite) into the well (if required). Have blowout-prevention equipment on site, should it be necessary for controlling the well.
- Geophysically log the well and survey it with a video camera prior to abandonment. The nature and extent of the geophysical logging depend on the value of the well and reason the well was abandoned. For those logs that can be effectively performed in the cased portions of the well, perform all geophysical logs from the bottom of the borehole to land surface. For all logs that can be performed only in open holes, run from the bottom of the borehole to the bottom of the casing.
- Fill the bottom of the borehole with clean, washed, and chlorinated (disinfected) gravel. Tap the gravel periodically to ensure placement.
- Place a bentonite/sand cap on top of the gravel at the top of the formation.
- Then fill the hole with neat cement grout from the bentonite/sand cap to land surface.
- Contain all fluid displaced during cementing and dispose of it in an environmentally safe manner.
- Submit a final well abandonment plan to the regulatory agencies at the time of application for a well abandonment permit, subject to field modification based on logging and conditions encountered during the abandonment.

Utilities should maintain records of abandoned wells and include the following information:

- Survey coordinates of the well
- Identification of markers or other indicators of abandoned well location
- The date the well was taken out of service
- The date the well was properly abandoned
- A description of the methods and materials used to abandon the well
- Reasons why the well was taken out of service
- Historic performance records of the abandoned well

If original records have been misplaced, contact manufacturers and well drillers, who maintain itemized records, including details for original pumps sold and installed. Operators should collect and incorporate these records into a permanent file system. These records provide original pumping levels, head conditions, and pumping conditions and can therefore indicate general area changes in water levels. Drilling logs may be replaced by running gamma-ray logs in both new and old wells and comparing them to correlate formation compositions (refer to Bloetscher et al. [2007] for more discussion of gamma-ray techniques). Changes in water quality may be documented in state health agency files and be relevant to well maintenance. Retroactive historic data of local aquifer conditions may also be available through records maintained by regulatory agencies.

REFERENCES

AWWA (American Water Works Association). 2014. Manual of Water Supply Practice M21, *Groundwater,* 4th ed. Denver, CO: AWWA.

AWWA. 2006. Standard A100, Water Wells. Denver, CO: AWWA.

Bloetscher, F., A. Muniz, and J. Largey. 2007. *Siting, Drilling, and Construction of Water Wells.* Denver, CO: AWWA.

Dania Beach. 2013. Copy of water-use permit file drawing for new pumps. City of Dania Beach, FL.

This page intentionally blank.

Chapter 7

Example ASR Programs in US

Within the United States, more than 70 ASR projects have been in operation for a decade or longer, a dozen of which have contributed to the understanding of alternative groundwater injection programs due to extensive discussion in the literature and at conferences. Some of these projects have provided successful service, while others have not lived up to their goals. Understanding both successful systems and those that have faced challenges is instructive. About a quarter of the ASR sites in the United States are not currently active, with clogging, inadequate recovery, and metal migration among the most common reasons (as defined in chapter 4). This chapter provides several examples of both operational and inactive ASR projects in a range of geographic and hydrologic settings.

The projects discussed here were chosen for two reasons: they are well known, or are projects that the authors have personal knowledge of that demonstrate particular issues. In each case, discussion of the project includes the following to the extent information is available:

- Background
- Project goals
- Analyses and results

PEACE RIVER ASR PROJECT

Background

The Peace River is the largest flowing surface water body in the area: a 2,480-mi^2 watershed that extends northeast for approximately 105 miles from Charlotte Harbor. The Peace River Manasota Regional Water Supply Authority provides the regional public water supply needs of Charlotte, DeSoto, Manatee, and Sarasota counties in southwest Florida. The Authority owns and operates a 24-mgd regional surface water supply and treatment facility on the Peace River. Near the intake structure of the Peace River Regional Water Supply Facility (PRRWSF), the river has an annual average flow of 970 mgd. Regulatory requirements limit river diversions for public supply to 10 percent of annual average flow, or up to 97 mgd.

Project Goals

The goal of the project is to store water for the dry season. The Peace River varies from high flows during the wet summer months to low flows during the dry winter and spring months. The Authority's permit to withdraw from the Peace River prohibits any diversion or withdrawal from the river when the flow is less than 84 mgd, a regular occurrence during the dry season. Because of this restriction, the Authority has pursued ASR to provide adequate water storage from the river during wet season flows, to access during the dry season when river flow is low and withdrawals are prohibited. ASR maximizes the amount of Peace River water that can be used and reduces the need for new supplies to be developed. The Authority has successfully pursued and operated a potable water ASR system for over 20 years.

Figure 7-1 shows a schematic of the existing Peace River water supply and treatment system. This shows that current facilities consist of a diversion structure, an off-channel reservoir, a surface water treatment plant, an ASR system, and above ground storage. Note that there are no impoundments on the Peace River; all seasonal storage must be provided by the off-channel reservoir and the ASR system.

Figure 7-2 is a cross section of the ASR system. Two operational storage zones are utilized at the facility. The upper storage zone is identified as the Tampa Limestone, located between 400 and 500 ft below land surface. Only one well is completed into this zone because of its relatively low water-producing ability and the proximity of nearby domestic wells. The major ASR zone is the Suwannee Limestone, located between 600 and 900 ft below land surface. The native water quality of this zone is nonpotable, with a total dissolved solids (TDS) concentration of approximately 800 mg/L. Twenty-one ASR wells (with yields of approximately 1.0 mgd each) currently are completed and operating in the Suwannee Limestone. An additional ASR storage zone is currently being tested in the Avon Park Limestone, a fractured limestone and dolomite formation beginning approximately

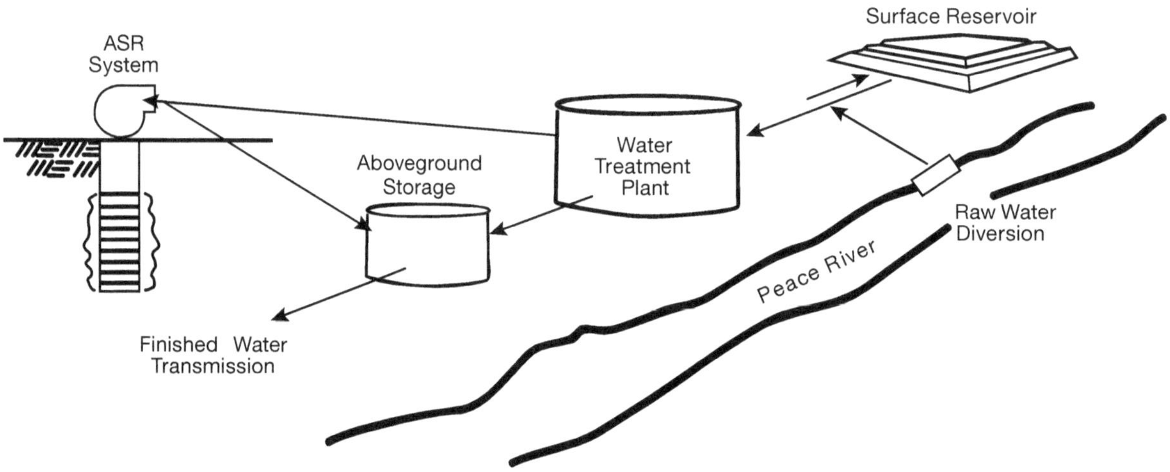

Figure 7-1 Schematic of Peace River ASR system

Source: *AWWA 2014*

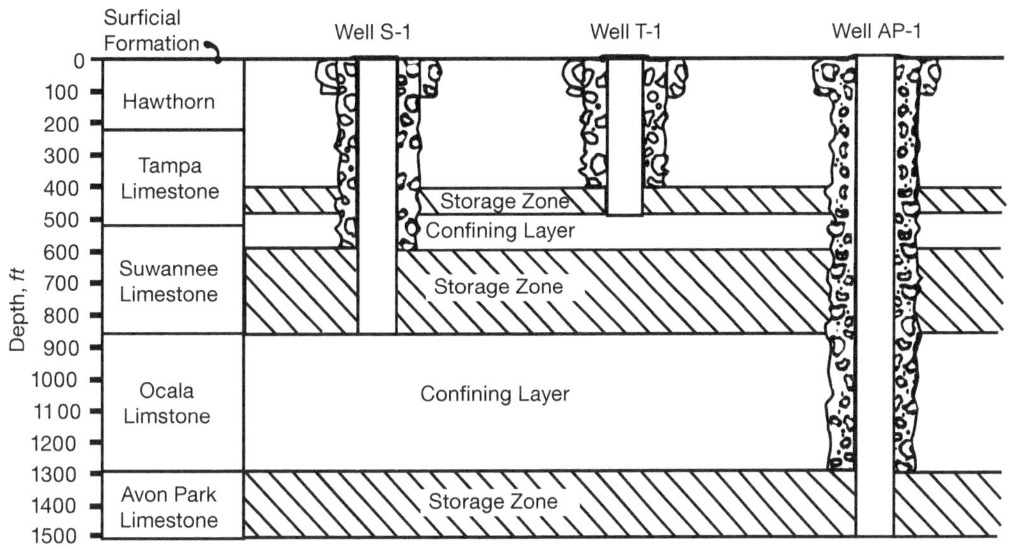

Figure 7-2 Cross section of the Peace River ASR system

Source: *AWWA 2014*

1,300 ft below land surface. The Avon Park Limestone is a high water-producing formation (well yields up to 3.0 mgd) with a TDS concentration of approximately 2,000 mg/L.

The PRRWSF's ASR wells are operated as needed to store treated drinking water when Peace River flow is available for withdrawal and system demand is less than the treatment capacity. The treated water is recovered from the ASR system when river flow is low and water demands are high. The operations staff monitors the ASR system for flow rates injected and recovered from the wells, cumulative volumes stored, the water quality of the injected water, and water quality of the recovered water. TDS is a general indicator of potable water quality and the parameter of primary interest for the water produced from the ASR wells at the PRRWSF. The TDS of the drinking water produced by treating Peace River water is typically 300 mg/L, which is well under the federal secondary drinking water standard for TDS of 500 mg/L.

Analyses and Results

The PRRWSF ASR system was designed to provide a six-month supply of water that meets drinking water quality standards. That quantity was determined through an analysis of historical flow records of the Peace River. The longest historical period in which no diversion from the river was allowed was seven months during a very severe drought in 1985. The ASR program has met project objectives and provided adequate water supplies during the winter and spring dry seasons. As a result, the system has continued to expand.

Currently, PRRWSF operates 21 wells and stores more than 1.6 billion gallons of treated drinking water. To meet the needs of a six-month dry season, each Suwannee limestone well must have 350 million gallons of stored water. The eight Suwannee Limestone wells each have a nominal capacity of 1.0 mgd. During a six-month recovery period, approximately 180 million gallons of water can be produced from each well. This means that approximately 50 percent of the stored water can be recovered at any one time to meet the available demands. Total recovery percentage increases over time from 50 percent to almost 100 percent of the water injected during successive storage and recovery cycles. Typically TDS concentration increases throughout the recovery period until the targeted volume of water from the ASR well has been withdrawn.

Lessons learned: ASR can work on a large scale, but the size of wells and design of the wellfield are critical. Freshwater aquifers can provide high recovery.

—*Study from AWWA 2014*

DENVER BASIN ASR PROJECT

Background

The principal water supply for the Denver Basin has been from the Arapahoe Aquifer, a bedrock aquifer lying below the service area. The Arapahoe Aquifer is one of the principal water-bearing units in the Denver Basin. The basin covers 6,700 mi^2 along the Colorado Front Range. The formation consists of 250-ft to 700-ft sequences of embedded limestone, sandstone, siltstones, and shale. At any point in time, the aquifer has seven to 10 distinct limestone units capable of producing commercially viable water supplies. Because of western water rights legislation, the embedded limestone units within the aquifer formation are considered one hydrologic unit, and wells are typically open in all units rather than specific ones.

However, these hydrologic units are not all identical with regard to aquifer characteristics and water production capability. The upper sandstone units are typically fine-grained and angular, while lower sandstone units tend to be more coarse-grained. The thickness of each unit tends to increase with depth, with the thickest sequences at the base of the aquifer where transmissivity may be lowest. The aquifer's thickness varies from 250 to 300 ft; the hydraulic conductivity is two to seven feet per day (ft/d), and transmissivity ranges between 4,000 and 16,000 gallons per day per foot (gpd/ft). The Arapahoe Aquifer in the area is approximately 600 ft thick, with the top lying approximately 1,000 ft below ground level. As a result, it is fully confined within the project area.

Historically, the Arapahoe Aquifer had artesian flows, dating back as early as the early 1800s. However, artesian pressure surface is now 500 to 600 ft below ground surface and continues to decline at an annual rate of 20 to 30 ft due to increasing water demands and the limited ability for this confined aquifer system to recharge. The base of the aquifer generally consists of an impermeable strata.

Several smaller utilities have attempted to use the Arapaho Aquifer for water supply instead of relying on surface waters like Denver Water, whose water supply comes from the South Platte River and flows from the Western Slope of the Continental Divide. While the state legislature has a 100-year plan to manage the aquifer, Denver Basin utilities are planning to mitigate impacts of withdrawal and drought long before the 100 years dry up the aquifer.

Project Goals

Utilities investigating ASR in the Denver Basin have done so because of

- declining artesian water levels in the Arapahoe Aquifer;
- the potential for excess surface water supplies in the South Platte River;
- improved management of water supplies;
- a degree of water supply security and drought resistance ensured by diversifying water supplies from ground, surface, and ASR projects; and
- optimization of existing storage and treatment facilities.

Consultants estimate that the Arapahoe Aquifer has 140,000 acre-ft of storage in the Denver Basin and the groundwater does not exceed any primary or secondary drinking water standards; therefore it can be piped directly into the distribution system after disinfection. The other aquifers in the area do not meet all drinking water standards and generally require treatment.

A series of ASR projects have been undertaken in the Denver Basin. Projects by Willows Water District (now East Cherry Creek Valley Water and Sanitation District) and Centennial are the most well-known. At the start of these projects, the engineers recommended that the two districts pursue a conjunctive-use plan to utilize their own water allocations, as well as purchase supplies from Denver Water. Because the summer demands can be as much as eight times those of winter demands, ASR was thought to be an ideal application for storing high South Platte River flows in the Arapahoe Aquifer.

The Willows project was a test program. Denver Water agreed to provide up to 500 acre-ft of water per year to the Willows Water District for the test program of injection and storage in the Arapahoe Aquifer. This water was treated at Denver's Foothills Water Treatment Plant and met all primary and secondary standards. The water was high in dissolved oxygen concentration and contained a chlorine residual. The Willows Water District's ASR Well 6 had a 19.3-in. diameter casing pipe and an 8-in. pump. A 400- to 1,200-gpm pump was specified. One monitoring well was installed 110 ft northwest of the proposed injection well to monitor the bubble formation (required by law). Water levels, water temperature, specific conductivity, dissolved oxygen, and pH were monitored throughout the injection cycle.

A series of injection cycles occurred during the winter season when surface and groundwater facilities were pumping minimal amounts, as shown in Figure 7-3. Limited withdrawal occurred. The primary goal for the first year was to pump short injection runs, as the focus was on hydraulic parameters related to the injection, water quality compatibility, and the effects of thermal variation between the water supply and the native water. In 1993, 175 acre-ft of water was injected, and the well had to be rehabilitated using carbon dioxide because of the presence of iron bacteria in the injection well, decreasing its capacity.

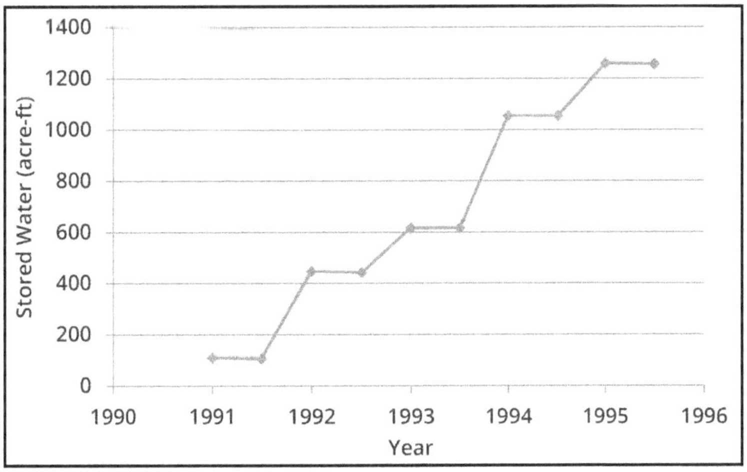

Figure 7-3 Injection and withdrawal cycles of Willow Water District's Well 6, 1991–1995

Analyses and Results

Results of the five-year testing program and the injection cycles were informative.

- The head buildup in the well during the injection process was greater than expected. Clogging and/or chemical reactions were initially suspected, but the difference in temperature between the injected water and the ambient Arapahoe Aquifer water led to changes in the viscosity of the water, a delay in mixing, and reduced injection rates.
- Drawdown increased during the pumping cycles as the project progressed. Fouling and clogging were observed. Injection rates were less than desired at the Willows system.
- Control of microbiological activity was required to prevent biofouling of the screen and gravel pack surrounding the well. High dissolved oxygen concentrations in the treated water from Denver Water contributed to microbiological growth in the gravel pack.
- Total trihalomethanes (TTHMs) in the water periodically exceeded the MCL (100 µg/L). Current data suggest that natural attenuation of TTHMs by microbiological activity within the aquifer formation during storage will sufficiently mitigate elevated TTHM occurrences. The potential to limit total organic carbon (TOC) and natural organic matter (NOM) in the source water could improve the process.

The majority of these factors can be controlled, excepting the thermal difference between the injected and native waters, which created a host of concerns with head drawdowns, specific conductivity, and transmissivity of the formation. No water quality problems or chemical reactions occurred that would be a concern for future injection programs. Because of the low performance this well was mothballed for a number of years and eventually sold to the East Cherry Creek Valley Water and Sanitation District.

Lessons learned: Disinfection by-products will decrease with time in ASR wells. However, the quantity of water injected and recovered from the aquifer may be limited to 1 mgd.

—Study from Pyne et al. 1996

COMPREHENSIVE EVERGLADES RESTORATION PLAN PILOT

Background

Water supply, water quality, and the Everglades' ecosystem health are intrinsically linked in south Florida. When attempting to evaluate the ecological health of southeast Florida, the entire southern portion of the peninsula of Florida must be considered. Historically, no barriers or canals existed to direct or control the path of the natural flow of water. The subsequent canal system, built in between 1930 and 1968, permanently reduced groundwater levels along the coast and enabled the development that exists today. However, the Biscayne Aquifer does not recharge as it once did because of those reduced groundwater levels, reduced historical flows to the Everglades, and less standing water in the Everglades during the summer months. The change in the land use along the coast has resulted in water falling on impermeable land where the water collects in pools or runs off rapidly where development has taken place, further reducing the potential for recharge.

Runoff from impermeable regions often results in large-scale flooding because of the storm intensity (rate of rainfall). The solution has been to use the canals to discharge this excess runoff to the ocean to minimize flooding. The net result is a reduction in available freshwater supplies during the dry season, which coincides with increased winter population and the peak irrigation season for lawns and agriculture. Subsequently, south Florida lacks adequate protection from recurrent drought during the drier winter and spring months and increased saltwater intrusion potential in coastal wellfields.

The South Florida Water Management District (SFWMD) is tasked with the management and protection of regional water resources, including competing interests of water quality, flood control, natural systems, and water supply. SFWMD's boundaries extend from central Florida to Lake Okeechobee, from coast to coast—Fort Myers to Fort Pierce—and south through the sprawling Everglades to the Florida Keys. Lake Okeechobee is the second largest natural lake within the continental United States and a principal water supply source for recharging the Everglades in the summer months. Lake Okeechobee was bermed as part of the central and south Florida Flood Control Programs that began in the 1930s, and today is operated as a completely managed system for water deliveries. SFWMD, as the system operator, in conjunction with Army Corps of Engineers, reviewed the potential to store the large quantities of water that are currently discharged to the Atlantic Ocean within an ASR system at Lake Okeechobee.

Project Goals

The goal of the Comprehensive Everglades Restoration Plan (CERP) ASR Project was to investigate the potential for the injection, storage, and recovery (after a period of time) of freshwater from the brackish lower Floridan Aquifer. The CERP included a plan for 330 ASR wells, at a capacity of 5 mgd each, to store water in the slightly brackish upper Floridan Aquifer for later retrieval and placement into the canal systems. Because of the surplus of freshwater in Lake Okeechobee and the Everglades system during the summer season and the lack of suitable water availability during the dry season, as well as the lack of reservoir areas, aquifer storage and recovery is an obvious candidate to be a part of the long-term water supply solution in south Florida. In addition, placement of the water below land surface eliminates the seepage and losses common to the natural system.

The Upper Floridan Aquifer formation occurs approximately 500 ft below land surface, beginning with what is called the Ocala Limestone. A confining zone is created with the Tampa Limestone and Hawthorn Formations, which exist throughout south Florida. The Ocala Limestone and the upper part of the Avon Park Formation constitute the upper Floridan Aquifer System. Because this aquifer is generally considered a future water supply aquifer, it is not used for injection. The lower part of the Avon Park Formation, which is the middle zone of the Floridan Aquifer, is generally considered a semi-confining unit and was bypassed. All the investigations were conducted on the Lower Floridan aquifer, which begins approximately 1,100 ft below land surface, with permeable zones within that formation occurring from 1,100 to 1,400 ft below land surface. Background water quality information gained for the Lower Floridan Aquifer indicated that chloride concentrations were 3,100 mg/L in the injection well and 2,200 mg/L in the associated monitoring wells. These zones are sufficiently confined from those above and below by intermediate dolomite units.

Wells at two sites were pursued for CERP. The first well tested was at Lake Okeechobee. The Lake Okeechobee Aquifer Storage and Recovery Project was undertaken by SFWMD in 1989 to assess the recoverability of injected canal water stored during the wet summer season for retrieval from the Lower Floridan Aquifer during the dry season. A series of starts and stops occurred with the proposed well testing program. The intent

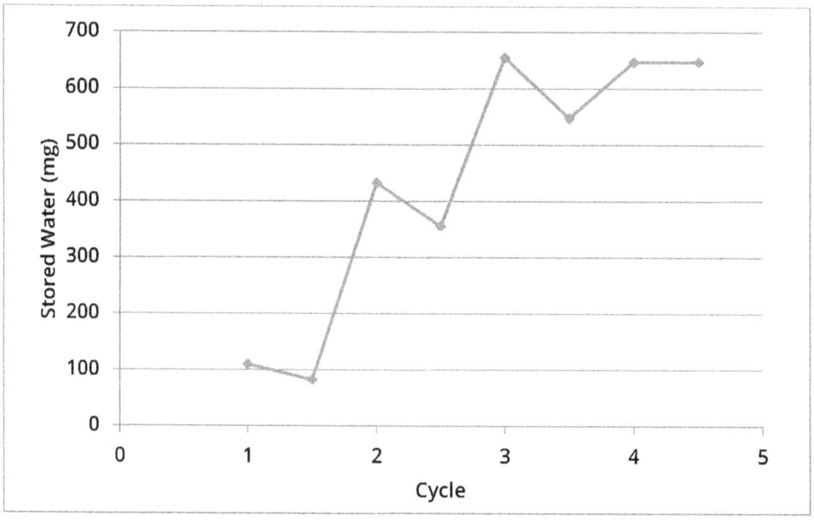

Figure 7-4 ASR injection and recovery cycles for the Lake Okeechobee Project

was to evaluate each hydraulic unit and pursue an ASR test project to determine the feasibility of large-scale ASR in the vicinity of the lake.

Analyses and Results

Four injection cycles were conducted for the Lake Okeechobee Project, each followed by recovery phases (Figure 7-4). The best recovery for a cycle was 36 percent. Injection volumes were 5.0 to 5.75 mgd. Storage time was limited to less than a month on all four cycles. The major result of the pilot project revealed that as more water was injected, the percentage of water that could be recovered increased. This is not unexpected, since the additional water was contributing to an added buffer zone between the native and the injected water. However, it was expected that density stratification would occur because of the lighter freshwater in the brackish water. Unknown at the time was what effect long-term storage would have on the stratification of the freshwater. Extensive efforts have been constructed for dealing with clogging and potential metals mobilization. The system is inactive according to the SFWMD, but the Army Corps of Engineers has completed its efforts and believes the well can be successfully converted to operations.

> *Lessons learned*: Large-scale ASR may be challenging and require time to resolve issues. Arsenic leaching may be an issue with limestone aquifers. Clogging from surface waters may need to be addressed. Geochemistry should be investigated.
>
> —*Study from Quinones-Aponte et al. 1996, updated by Mirecki 2014*

COLLIER COUNTY, FLORIDA, PROJECT

Background

Collier County is located in southwest Florida and is fortunate to have abundant water supplies. However, the area is subject to cyclical subtropical climatic patterns that create distinctive wet and dry seasons. Most of the rainfall accumulation that does not evapotranspire drains to the Gulf of Mexico through an extensive canal system without replenishing environmental needs. The result is a reduction in available freshwater during the dry season, which coincides with the increased winter population and the peak irrigation season for lawns and agriculture.

Collier County has been among the fastest growing areas of the United States since 1980. To meet these future demands, the Collier County Water–Sewer District explored a number of methods to expand treatment capacity and store large quantities of water for peak periods, thus eliminating plant expansions intended solely to meet these peak demands, including ASR in the early 1990s.

Project Goals

The aim of Collier County's aquifer ASR system was to pump potable water into the Hawthorn Aquifer zone, between 465 and 525 ft below the surface, and retrieve it during peak periods. The Hawthorn zone is a saline aquifer, with a TDS concentration of approximately 5,000 mg/L. This creates enough differential in density that the fresh, potable water will try to stratify, forming the freshwater bubble in the upper portions of the aquifer. The potable water will later be retrieved, rechlorinated, blended in a storage tank, and pumped into the system.

A 1,600-ft test well was first drilled to determine the most favorable aquifer for the ASR project and to determine native water quality, aquifer parameters, and confinement parameters of the units drilled. Drilling of the ASR well and three monitoring wells occurred in 1991 and testing began later that year. Seven injection, storage, and recovery cycles varying from 4 to 100 million gallons each have occurred in the first five years. Initial injection cycles were limited, providing the staff and hydrogeologists with the ability to observe the movement of freshwater by analyzing the chloride content of the monitoring wells. Several storage cycles lasted over six months, allowing stratification to occur in the aquifer and providing a mechanism for estimating the amount of water that would be permanently lost to mixing. Testing showed that with each succeeding cycle, the rate of chloride concentration increase per unit volume of water recovered decreased significantly. This relationship indicated that each test cycle improved the recovery efficiency of the ASR system. Recovery was limited to about 300 mg/L chlorides as this is the maximum concentration to be blended with the 60 mg/L chloride content of the potable system water.

Analyses and Results

Initial cycles showed low freshwater recovery amounts because of the small amount of water injected. Later cycles showed higher return as the volume of injected water increased. An estimated 300 million gallons of water or more are necessary to create and maintain the saltwater-to-potable-water transition zone in the formation. After that point, recovery returns were expected to exceed 85 percent. The Collier ASR well is the only well to test the buoyancy effects of the bubble in a brackish aquifer over a 6-month period. The bubble did float to the upper half of the 40-ft formation within 180 days, impeding recovery.

As a result, this project showed that utilities considering ASR technology should understand that the initial transition zone from freshwater to saltwater may require up to 500 million gallons of water to obtain a reasonable recovery rate. This requires that the utility be of significant size and have available capacity during wet periods to inject a minimum of 1.0 mgd into the ASR zone. The size of the transition zone, 300–500 million gallons, was first determined at the Collier County facility.

The facility was also the first in Florida to use distribution pressure to inject potable water, was among the few to utilize a saline aquifer for storage, and was the first to attempt to pump the retrieved water into the distribution system without retreatment other than disinfection. Chemical analyses indicated that no problems were expected to develop with this concept.

The project confirms the suppositions that ASR allows treatment plants to be sized for average demands, rather than seasonal high demands, and thereby can save capital infrastructure costs. Additionally, Collier County's ASR system is located near the end of the distribution system, which eliminates the need to construct a larger water main from the treatment plant to meet peak demand periods.

> *Lessons learned*: In abrackish aquifer, the freshwater bubble will float quickly. In a 40-ft-thick formation the bubble migrated to the top half in less than 6 months. Multiple monitoring wells can be used to triangulate the location of the bubble. The transition bubble may be large, 500 million gallons or more, and take years to develop.
>
> —*Study from Bloetscher et al. 1996*

MEMPHIS, TENNESSEE, PROJECT

Background

The Memphis Light, Gas and Water Division (MLGW) is a municipal public utility serving the City of Memphis and Shelby County, Tennessee. MLGW serves 420,000 customers and is owned by the city. Since 1939 the MLGW has provided electricity, natural gas, and water service for residents of Memphis and Shelby County. MLGW relies on an aquifer that sits beneath Shelby County to supply its water. MLGW delivers water to more than 257,000 customers through more than 175 wells throughout Shelby County. MLGW reports that the Memphis Aquifer is one of the largest artesian well systems in the world.

Project Goals

The goal of the ASR project was the storage of treated river water for later use. The project was initially designed for operation, but the goals shifted to a proof of concept project. Water quality issues were a major interest.

Analyses and Results

MLGW tested an ASR project from 1996 to 2000. The source water was treated water from the Memphis municipal groundwater supply. The recovered injectate was aerated, filtered, chemically stabilized, and disinfected prior to distribution and met all requirements of the Safe Drinking Water Act. The well was put through five cycles of injection, storage, and recovery. Within the first few months of recovery, high THMs were observed in the recovered water.

The injection was into a freshwater aquifer, so recovery volumes were high. The THM values in the recovered water were typical of the treated distribution water. MLGW dedicated the fourth and fifth cycles strictly to solving the problem of the high THMs. Because the THMs were persistent in the recovered water, MLGW used hydrogen peroxide to remove the residual chlorine in the water before injection on these cycles. The hydrogen peroxide proved to be an effective dechlorination agent and eliminated the THM-formation problem.

However, according to the Tennessee Underground Injection Control program, the ASR project was halted by MLGW's Water Quality Assurance Laboratory because of the lack of indisputable evidence that no harmful by-products were produced as a result of the hydrogen peroxide pretreatment. The termination of the project occurred even though the laboratory tests failed to identify any potentially harmful compounds in the water. The project has been discontinued and there are no plans to resurrect it, according to the state agency that permitted it 25 years ago.

> *Lessons learned:* This project might have been successful had the data from Colorado's Willows Water District project been available at the time. Water quality issues like trihalomethanes may become regulatory agency issues. Discussion with the agencies to address chlorine in injectate is needed at the start of the project.

HORRY COUNTY, SOUTH CAROLINA, PROJECTS

Background

Horry County has two ASR well exploration projects. Both are tested wells in the Black Creek Aquifer, a sandstone formation, 300 ft to 600 ft deep, that has been the water supply for much of the eastern Carolinas since the 1940s. The aquifer does not recharge appreciably with time as evidenced by large aquifer declines that have effectively depleted it in both North and South Carolina. Myrtle Beach, South Carolina, relied on groundwater until 1990, but the drawdown and water quality concerns (fluoride concentrations exceed the MCL of 4 mg/L) required that the city convert its system to a surface water source.

The Myrtle Beach ASR well investigation was conducted by the US Geological Survey (USGS) between 1989 and 1994 using an old water supply well to meet peak summer demands that are double the average day demands. The city had limited storage for summer peaks.

The Grand Strand Water and Sewer Authority (GSWSA) serves Horry County, home of Myrtle Beach. GSWSA also converted its system to a surface water source in 1990. Its wells are also screened 300 ft to 600 ft deep in the Black Creek Aquifer. The first well was placed into operation in 1996. GSWSA used its existing production wells—which were to be abandoned—as its ASR wells. The initial investigations into ASR in the area were conducted by the USGS at the same time as the Myrtle Beach well investigation.

Project Goals

In both cases the utilities desired to store water for use during the peak summer season. Both had recently converted from groundwater to surface water as a result of declining well yields. Because the period of high demand is relatively short, project planners determined that building peak capacity plants was not cost effective as they would be underutilized. Other utilities in coastal South Carolina also determined that it was not practical to build treatment plants to meet peak demands.

Analyses and Results

To make the system viable, Myrtle Beach wanted a system where the withdrawn water was to be treated by disinfection only. However, the well could not meet this requirement as extensive carbon dioxide appeared in the recovered water and could not easily be removed without additional treatment for degassing prior to introduction into the distribution system. The investigation also indicated that piping costs to replumb the system to incorporate ASR were higher than anticipated. The uncertainty and potential costs associated with these modifications caused Myrtle Beach to discontinue ASR as a water supply strategy, deactivate the project, and ultimately sell its system to GSWSA.

GSWSA continued to pursue ASR because of high peak demands—17-mgd average daily demand versus 26-mgd peak demand. Recovered water is disinfected and introduced into the distribution system. The carbon dioxide problem that plagued Myrtle Beach does not appear to be a problem for GSWSA. The wells inject an average of 300 to 500 gpm and withdraw 500 to 700 gpm. Currently, GSWSA has the most extensive ASR system in the Carolinas, with 20 wells.

> *Lessons learned:* ASR can work on a large scale when it involves many smaller wells. Freshwater aquifers can provide high recovery. Use of old water supply wells may provide a good option to pursue ASR while reducing costs.

ROSEVILLE, CALIFORNIA

Background

The city of Roseville is located in California's Central Valley near the state capital Sacramento. Roseville's primary water supply source is surface water from the Folsom Reservoir on the American River. The reservoir is filled with water from two sources: a water contract with the Central Valley Project for 32,000 acre-ft annually (AFA), along with 30,000 AFA purchased water rights from Placer County Water Agency. In 1997, the utility supplied water to 67,000 people (Petersen and Glotsbach 2005). The city's water needs have been steadily rising, with deliveries increasing 50 percent from 1990 to 1996 and expected to increase 150 percent by buildout (City of Roseville 1992).

Project Goals

The city is studying the feasibility of ASR using dual purpose (injection/extraction) wells to sustain water supply reliability in one of the fastest growing regions of the state (Petersen and Glotsbach 2005). With more than half of all Californians relying on groundwater for at least a portion of their water supply, groundwater supplies approximately 30 percent of urban and agricultural demands. In drought years, this percentage increases to more than 40 percent (DWR 2003).

The intent of the ASR project was to inject potable water into the aquifer. The planned citywide ASR would be the first implemented in the Central Valley. The North American sub-basin lies in the eastern central portion of the Sacramento Groundwater Basin and covers 548 mi^2 of surface area in Sutter, Placer, and Sacramento counties. The basin is bounded by the Bear River to the north, the Feather and Sacramento Rivers to the west, and the American River to the south. As identified by DWR Bulletin 118 (2003), the eastern boundary is defined by a north–south line extending from the Bear River to Folsom Lake.

Analyses and Results

The pilot study was conducted from May to September 2004. Baseline monitoring and sampling included a series of monitoring and sampling events performed from May 5 to June 16, 2004. During this period, no water was injected at the Diamond Creek Well. The purpose of baseline monitoring and sampling was to establish the groundwater elevations and water quality in the aquifer before injection (Petersen and Glotsbach 2005).

The injection portion of the cycle test consisted of 26 days of continuous surface water injection followed by two days of rest, or storage (Petersen and Glotsbach 2005). A total of 51 million gallons of water was injected. The testing of the injection cycle included four monitoring and sampling events, performed from June 16 through July 12, 2004, to determine changes in groundwater elevations and water quality of the aquifer at various distances from the injection well during and following the injection period. System pressure and injection flow rate were also monitored during the injection phase of the pilot test.

There were three extraction phases during the study: July 14 through 26, August 3 through 9, and September 9 through 20, 2004 (Petersen and Glotsbach 2005). The three extraction phases were separated by storage periods lasting 8 and 31 days. The total volume of water extracted during all three phases was 143 million gallons. The volume extracted substantially exceeded the injected volume to ensure that essentially all of the injected water was removed from the aquifer.

The project planners recognized that THMs and HAAs were introduced into the aquifer with the injected water, at trace concentrations, and they anticipated that the DBPs might increase slightly during the storage phase of cycle testing as residual chlorine in the potable water continued to react with the DOC in the potable water. This was confirmed, but the DBP concentrations decreased very quickly during the storage and extraction portions of the short-term pilot test (Petersen and Glotsbach 2005).

After injecting water into the aquifer from December 2005 to May 2006, the city began extracting drinking water from the well and delivering it to Roseville water customers beginning in July 2007. This extraction was done in accordance with the regulatory requirements placed on the city's use permit.

The long-term goal of Roseville's ASR program is to implement a full-scale project of up to 12 wells, capable of injecting 10,000 acre-ft of water per year. The ASR wells would help meet operational needs to ensure that water reliability of the system is maintained (City of Roseville 2014).

> *Lessons learned*: When implementing an ASR project, consider potential technical, regulatory, and operational challenges. ASR can provide flexibility and resilience in the water supply. DBPs are a potential issue for regulatory agencies.

CONCLUSIONS

ASR is certainly a concept that will play an important role in the solution to water supply concerns, not only for potable and residential users, but also for agricultural, industrial, and environmental users. ASR is a viable method (subject to proper geologic conditions) for improved water management. Regulators will need to be flexible in allowing demonstration projects to proceed in a cost-effective manner to enable full understanding and possible assessment of the role of ASR in water supply solutions.

Much can be concluded from historical lessons learned from existing and ongoing ASR projects. There are clear examples of successful programs, as well as examples of projects that were not successful. ASR programs need more care and planning than just pumping water in and taking water out. Client expectations, geology, water availability, and other issues drive success. Additional research is needed to better describe scenarios where ASR designs experienced unique challenges.

The question now is, what can be done to improve our understanding of the hydrogeological setting to allow safe development of the ASR concept on a broader scale? While ASR appears to be moving forward as a major component of water supply management, much more information is needed to definitively assess its value and possible environmental impacts.

REFERENCES

AWWA (American Water Works Association). 2002. Manual of Water Supply Practice M21, *Groundwater,* 3rd ed. Denver CO: AWWA.

AWWA. 2014. Manual of Water Supply Practice M21, *Groundwater,* 4th ed. Denver CO: AWWA.

Bloetscher, Frederick, Albert Muniz, and Gerhardt M. Witt, 2005. *Groundwater Injection: Modeling, Risks, and Regulations.* New York: McGraw-Hill.

Bloetscher, Frederick, Charles W. Walker, and W. Kirk Martin, and Thomas M. Missimer. 1996. Aquifer storage and recovery to meet peak water demands in southwest Florida. Proceedings of CONSERV96, pp. 825–829. Denver, CO: AWWA.

California Department of Water Resources (DWR). 2003. *California's Groundwater.* Bulletin 118 – Update 2003, October. State of California.

Petersen, Christian E., and Kenneth Glotzbach. 2005. Aquifer storage and recovery for the city of Roseville: A conjunctive use pilot project. WEFTEC 2005 Proceedings. Alexandria, VA: WEF.

Pyne, R.D.G., P.C. Singer, and C.T. Miller. 1995. Aquifer storage recovery of treated drinking water. Denver, CO: AWWA Research Foundation.

Quinones-Aponte, Vincente, Kevin Kotun, and Joseph F. Whitly. 1996. Analysis of tests of subsurface injection, storage and recovery of freshwater in the Lower Floridan aquifer, Okeechobee County, Florida. US Geological Survey, Open File Report 95-765.

Roseville. 2008. Background on Aquifer Storage and Recovery Program and the Diamond Creek Well—Updated Jan. 2, 2008. www.waterboards.ca.gov/rwqcb5/board_decisions/tentative_orders/0508/aquifer_storage_recovery/roseville-asr-buff.pdf. Accessed 2/5/2014.

Roseville, City of. 1992. Roseville 2010 General Plan Final Environmental Impact Report.

Chapter 8

ASR Versus Other Groundwater Recharge and Storage Programs

Beneficially capturing water resources that are on hand now for use later is a common driving force behind aquifer storage programs. One option is aquifer storage and recovery, or ASR, which involves the management of water supplies for both potable (drinking water) and nonpotable water supply systems. While the purpose of aquifer reclamation and aquifer recharge is to sustain long-term demands in an existing or potential source, ASR alone is intended to store water until such time as is needed to meet peak needs, long-term growth demands, emergency conditions, or because of poor water quality. More than 200 sites in 27 different states have either utilized or investigated the concept of ASR.

Because of varying cost and political factors and local/regional aquifer conditions, ASR programs may be supplemented or enhanced via alternative technologies. Alternative recharge program goals may be to recharge aquifer systems for long-term supplies and groundwater availability, or to create hydraulic barriers to low-quality groundwater or seawater intrusion into fresh groundwater formations. The alternative approaches can include passive or managed aquifer recharge via spreading basins or injection wells. Regardless of the approach, it is key that any of these programs determine the capacities and limitations of the aquifer system, the groundwater chemistry, and potential effects through short- and long-term employment of the proposed storage or enhancement system.

While the goal of this manual is to focus on ASR technology, chapter 1 outlined other categories of recharge programs, namely artificial aquifer creation, aquifer recharge, and aquifer reclamation. At present, all of these strategies are proven, cost-effective programs capable of successfully using existing waters and sources that are managed for long-term water supply.

Aquifer recharge programs involve large recharge basins, such as those in Arizona, California, and Utah. The concept is to flood the basins with water and let it percolate into the ground. Washington County (St. George), Utah, developed this type of system to resolve long-term water supply problems, using surface water from the Virgin River. In Arizona, both surface waters and reclaimed wastewater are used as recharge sources. California primarily uses surface water sources when available. Recharge basins differ in two respects from ASR wells: They have the potential for large evapotranspiration losses and they assume a surficial aquifer. Most ASR systems involve confined aquifers.

Aquifer reclamation is a technique where large quantities of higher quality waters are injected into a depleted or a contaminated aquifer. Another application is the injection of freshwater into aquifer zones contaminated by saltwater. When an aquifer head is reduced by development or drainage, saltwater can continue to move inland and contaminate the aquifer formation. Because this movement is slow, proper injection procedures can push the isochlor line back toward the ocean by increasing the aquifer head. Injecting a source of freshwater may stabilize the salinity front or force it to retreat. This technique can be applied in coastal areas where saltwater intrusion has occurred. The Orange County and West Coast Basin Barrier projects in California are examples of this type of project. Artificial recharge with pilot wells was instituted in Pembroke Pines and Hollywood, Florida.

Chapter 11 in Manual M21, *Groundwater*, has more details and expanded explanations for these projects. To address complexities and study cases to date, the topic of managed recharge projects is expected to be further addressed in a future AWWA manual.

Appendix A

US ASR Systems

State	Project, Location	Estimated Start Year	Number of Wells in Project	Source of Water	Use of Water	Storage Cycles	Injection Cap (*mgd*)	Withdrawal Capacity (*mgd*)	Ratio In, Out	Peak Flow on Site (*mgd*)	Amount of Water Stored (*mg*)	Depth of Casing (*ft*)	Depth of well (*ft*)	Injection Horizon (*ft*)	Casing Material	Casing Diameter (*in.*)	Tubing and Packer?	Formation type	Transmissivity (*gpd, ft*)	TDS (*mg, L*)	Confinement
AZ	Recreation Centers of Sun City West	1999	5	reclaimed	irrigation	>10	1.7	1.7	1.0	8	1,800	1,000	1,000	n/a	S[1]	12	n	alluvial			
AZ	Cave Creek	2007	2	reclaimed	irrigation	<10	1.5	1.5	1.0	7.9	3,000	180	180	n/a	PVC[2]	12	n	alluvial			
AZ	Chandler Ocatillo	2002	4	reclaimed	irrigation	>20	10	10	1.0	20	3,640	150	240	90	S	18	n	alluvial	7,000	2010	sand, clay
AZ	Chandler Tumbleweed	1999	17	raw surface	raw	>20	1.5	1	1.5	10	1,800	39	350	311	S, SS[3]	18	n	alluvial	25,000	2000	
AZ	Chandler Intel	2005	4	river water	cooling	<5					3,100							alluvial			
AZ	Far West W&S	2009	2	reclaimed	irrigation	<10	1	1	1.0	2		80	180	100	PVC	12	n	alluvial			
AZ	Fountain Hills	2001	4	reclaimed	irrigation	>20	0.4	0.2–0.6		0.8	780	550	750	200	SS	18	n	alluvial	20,000 to 40,000		sand, clay
AZ	Gilbert	2005	2	reclaimed RO[4]	irrigation	>10	2	2	1.0	2	1,300	960	960	n/a	S	15	n	alluvial			
AZ	Glendale Arrowhead Ranch	2001	8	raw surface	raw	>10	1.5	1.5	1.0	3.6	750	630; 1,289; 1,290; 1,710	2,066; 1,540; 1,317; 1,725		S	21	n	alluvial			
AZ	Lake Pleasant Sewer Company	2007	3	reclaimed	irrigation		0.1	0.1	1.0	0.1	20		75	n/a		8		alluvial			
AZ	North Scottsdale	2006	3	raw surface	raw	<10	0.5	0.5	1.0	unknown		770	770	n/a	S	7	n	alluvial			
AZ	Peoria – Stone Street	1963	1	raw surface	raw	>10	1.6	1.6	1.0	1.6	31,525	1,129	1,129	n/a	S	16	n	alluvial			
AZ	Phoenix Water ASR	2007	1	surface	PWS[5]	<5	3.4	6.5	0.5	6.5	300	1,400	1,400	n/a	S, SS		n	alluvial	200,000		clay
AZ	Salt River Project	1995	8	raw surface	raw	<5	1	1	1.0	1	16,000	370	370	n/a	PVC	12	n	alluvial			
AZ	Sun Lake City – Pima Utilities	1998	5	reclaimed	golf course irrigation	>10	0.3	0.3	1.0	0.6	237	218	228	10	SS	14,16	n	alluvial			
AZ	Scottsdale	1999	5	raw surface	raw	>10	0.4	0.6	0.7	13	5,000	180	600	420	PVC	10		alluvial			

AZ	Tempe	2005	1	reclaimed	recharge	<10	0.5	0.5	1.0	0.5		440	450	10	S	11	n	alluvial			
AZ	Tormanto	2004	2	raw surface	raw	<10	2.5	2.5	1.0		628	175	175	n/a	PVC	16	n	alluvial			
AZ	Yuba City	2013	0	reclaimed	irrigation	0															
CA	Antelope Valley Los Angeles PWD	1994	2	groundwater	raw	3	1	1	1.0	1	40	280	710	430	S		n	alluvial	61,000	300	silt
CA	Boron CSD Well 15	2006	1	surface water	PWS	0	0	0		0	3,000	430	504	74	S perforated	14	n	alluvial			silt, clay
CA	Calleguas Municipal Water District	1993	18	surface water	drinking water	>10	0.8	1.5	0.5	4.5	250	800	1,200	400	S	18	n	alluvial	145,000	195	
CA	Camarillo	1979	2	surface water	raw groundwater	0	0	0		0	0										
CA	Carpinteria Valley	2004	1	surface water	raw	1	0.6	0.6	1.0	0.6	3.6	1,210	1,210	n/a	SS	14	n	alluvial	22,375		clay
CA	East Bay Municipal Utility District	1997	1	surface water	raw	<5	.13	.13		1	5.2	158	198	40	S	18	n				
CA	Eastern Municipal Water District, Perris	1994	1	surface water	PWS	3	1.3	1.1	1.2	1.1	0	620	620		S	18	n	alluvial			
CA	E Palo Alto	2012	1	surface water	PWS	1	0.5	0.5	1.0	0.5	1,500	259; 316	282; 325		S		n	alluvial	2,600	860	clay
CA	East Bay Municipal Utility District – Bayside	2009	1	ground and surface water	PWS	<5	1	2	0.5	2	1	653	653		S, SS		n	alluvial	75,000		clay
CA	Elsinore Valley Municipal Water District	2005	8	surface water	PWS	2	2 to 3	3 to 3		16	730	420; 340; 331; 640; 570	1,410; 1,260; 1,980; 1,660; 960	500–1,000	S with screen	16; 20; 32	n	alluvial			alluvial
CA	Fresno	2013	0	surface water	PWS	0												alluvial			clay
CA	Goleta Water District	1979	14	surface water	PWS	>10	1.1	0.4		0.4	375	300	1,300	1,000	SS			alluvial	130 – 2,000		
CA	City of Long Beach WMD	2002	4	surface water	PWS	>10	1.5	3.5	0.4	7	4,200	1,012	1,012		S	16	n	alluvial	72,000 – 150,000		clay
CA	Monte Vista Water District	2005	4	surface water	raw	>10	3	4	0.8		1,300	930	1,230								
CA	Monterey Peninsula WMD	2005	3	surface water	raw	>10	0.8	1.5	0.5	3	800	480	720			18		sandstone			

(continued)

State	Project, Location	Estimated Start Year	Number of Wells in Project	Source of Water	Use of Water	Storage Cycles	Injection Cap (*mgd*)	Withdrawal Capacity (*mgd*)	Ratio In, Out	Peak Flow on Site (*mgd*)	Amount of Water Stored (*mg*)	Depth of Casing (*ft*)	Depth of well (*ft*)	Injection Horizon (*ft*)	Casing Material	Casing Diameter (*in.*)	Tubing and Packer?	Formation type	Transmissivity (*gpd, ft*)	TDS (*mg, L*)	Confinement
CA	City of Oxnard	2003	5	surface water	PWS	>5	9	9	1.0	11	3,800	360	430		SS	18		alluvial			
CA	Palo Alto	1978	1	surface water	PWS							20	46	26	PVC	8	n	alluvial	80,000		
CA	City of Pasadena	1992	3	surface water	fire	>5	1.2	1.5	0.8	3	1,200	192	629	160	S	26	n	alluvial	100,000 –200,000		
CA	Pleasanton, Zone 7 Livermore	1998	4	surface water	PWS		1.7	3	0.6	5.7	1,000	216	490		SS	18	y	alluvial	140,000	600	clay
CA	Roseville	2005	2	surface water	PWS	>5	2	3		4	1,100	310	450	140	S		n	alluvial		450	
CA	Jurupa	2014	0	surface water	PWS																
CA	Santa Barbara, City of	1990	2	surface water	PWS	>10	0.5 to 0.7	0.6, 0.9		1.5	5,350	270 & 278	480 & 590		S		n	sandstone	7500		
CA	Santa Clara County Water District, San Jose	1999	1	surface water	PWS	5	1.5	0.6	2.5	0.6	70	250	560	300	SS		n	alluvial	140,000		
CA	Sonoma County Water Agency – Santa Rosa ASR	2004	1	surface water	fire	<4	0.8	2	0.4	2	14	400	800		S	16 and 20	n	alluvial			
CA	City of Tracy	2004	2	surface water	raw		1.5	3	0.5	3	100	320	770		SS	18	n	alluvial	98,000		
CA	Valley Water Company Foothills, Monk Hill Basin	1994	2	surface water	drinking water	>20	1	1.5	0.7	3	800	460 & 599	460 & 599		S		n	alluvial	6,100	<500	clay
CA	Victorville – Victor Valley Water District	2006	4	surface water	cooling	>20	0.8	1.2	0.7	4.5	1.5	550	800		epoxy SS	18	n	alluvial			none
CO	Castle Pines North AR4	1999	1	surface water	PWS	<5	0.5	1	0.5	1	222	1,457	1,750	293	S	12	n	sandstone	11,100		clay
CO	Centennial Water & Sanitation District, Highlands Ranch	1990	26	surface water	PWS	>10	0.5	0.7	0.7	6.7	4,500	922	1354	200	PVC, SS	10	n	sandstone	8,500	180	

CO	Colorado Springs Utilities Park Forest, Donata	2004	2	surface water	PWS	<5	0.7 & 0.1	0.8 & 0.1		5.5	2,300	250 & 650 & 1,225	450 & 1,100 & 1,700	200	S	5.5	n	sandstone	0.65	180	
CO	Consolidated Mutual Water	2004	6	surface water	PWS	>10	1.4	1.4	1.0	1.4	730	410	700	290	PVC, SS	12	n	sandstone			silt
CO	East Cherry Creek Valley Water District and Sanitation, Willows Water District	1990	4	surface water	PWS	>35		0.5-1.2		1.8	4,170	1,300	1,538	338	S	19	n	limestone			shale
CO	Parker Water and Sanitation District	1996	1	surface water	PWS	<10						250 & 650	450 & 1,100	650	S	5.5	n	sandstone	1.2	250	shale
CO	So Metro Water Supply Authority	2010	0	surface water	PSW													sandstone	9,000	900	shale
DE	United Water Delaware	2008	1	PWS groundwater	PWS	>5	0.3	0.3	1.0	1	225	567	675	91	S, SS	12	n	sand		150	clay
DE	Wilmington Manor Gardens and Artisans Village Sites	1998	1	PWS groundwater	PWS		0.4			1.3	1,000										
FL	Bonita Springs San Carlos	2000	1	drinking water	drinking water	>1	1	1	1.0	1	138	650	700	50	S		n	limestone		300	limestone, clay
FL	City of Boynton Beach,	1993	1	PWS groundwater	drinking water	>20	0.8–1.5	6.34		6.34	200	804	909	105	S	16	n	limestone	70,000	3900	clay
FL	Bradenton – Reservoir	2007	1	surface water	raw	6	3	3	1.0	3	150		432				n	limestone			limestone, clay
FL	Bradenton High Service Pumping Station	2003	1	surface water	drinking water	>10	1.5	1.5	1.0	1.5	140	465	505		PVC	17.4	n	limestone		1200	limestone, clay
FL	Broward County	1998	1	raw groundwater	raw water	5	0.8–1.6	6.34		6.34	1,000	995	1,200	205	S	16	n	limestone		3000	limestone
FL	CERP Hillsboro Basin	2001	1	surface	raw water	5	5	5	1.0	5		1,015	1,270	255	PVC	12	n	limestone			dolomite, clay
FL	CERP Lake Okeechobee	2004	1	surface water	canal recharge	<5	9.4	9.4	1.0	9.4		1,268	1,700	432	S	24	n	limestone		6000	dolomite
FL	CERP St Lucie	2001	1	raw groundwater	raw	4	5	5	1.0	5	1,800	600	770	170	PVC	12	n	limestone		1200	
FL	CERP Taylor Creek	1989	1	surface water	raw	74	0.3–0.5	2		2	1,000	1,268	1,710	442	S	24		limestone		4,000	clay
FL	Cocoa	1987	10	surface water	drinking water	>20	0.8–1.2	1		23	1,000	387	429	42	S	16	n	limestone			dolomite

(continued)

State	Project, Location	Estimated Start Year	Number of Wells in Project	Source of Water	Use of Water	Storage Cycles	Injection Cap *(mgd)*	Withdrawal Capacity *(mgd)*	Ratio In, Out	Peak Flow on Site *(mgd)*	Amount of Water Stored *(mg)*	Depth of Casing *(ft)*	Depth of well *(ft)*	Injection Horizon *(ft)*	Casing Material	Casing Diameter *(in.)*	Tubing and Packer?	Formation type	Transmissivity *(gpd, ft)*	TDS *(mg, L)*	Confinement
FL	Collier County North	2006	1	reclaimed	irrigation	<5	1	1	1.0	1.5		806	880	74	FG[6]	10	n	limestone		1,900	clay
FL	Collier County, Manatee Rd	1998	1	groundwater	surface water	>10	1	2	0.5	2	>80,000	465	528	63	PVC	12	n	limestone		5,000	clay
FL	City of Deerfield Beach	2001	1	groundwater	drinking water	<5	1.3	3.3	0.4	3		960	1,128	168	S , PVC	26	y	limestone	24,000	6,000	clay
FL	Deland	2009	1	groundwater	uncertain	3	1	1	1.0	1	20	880	1,087	207	PVC	5.625		limestone			
FL	Delray Beach	1996	1	PWS groundwater	drinking water	<5	0.4	0.4	1.0	3.7	50	1,016	1,120	104	S	14	y	sandstone		3240	clay
FL	Destin	2011	1	reclaimed	irrigation	>5	3	3	1.0	3	20	106, 110	156		PVC, SS	8	n	Sand and gravel			sand and gravel
FL	Englewood	1998	1	reclaimed	irrigation	4	2.2	1	2.2	1	700	512	700	188	S	16		carbonate	20,000		dolomite
FL	Fort Lauderdale Five Ash	1999	1	raw groundwater	raw water	3	0.5	2	0.3	2	<400	1,055	1,200	145	S	16	n	limestone		780	limestone
FL	Ft. Myers Winkler	1998	1	surface water	drinking water	>4	1	1	1.0	1	0	455	575	120			n	limestone		1770	limestone
FL	Hialeah	1976	1	raw groundwater	raw	3	1.3	5	0.3	5	320	955	1,105	150	S		n	limestone	1,200	4000	clay
FL	Lee County, Corkscrew	1996	6	groundwater	PWS	3	0.65	0.64	1.0	3.9		382	397	15	S	12	n	limestone		386	clay
FL	Lee County, North Reservoir	2000	1	surface water	raw	1	1	1	1.0	1		540	642	102			n	limestone		700	clay
FL	Lee County, Olga WTP	2001	1	groundwater	PWS	3	1	1	1.0	1		859	920	61			n	limestone		2700	clay
FL	Manatee County – Lake Man	1983	6	surface water	PWS	>20	2.1	0.2–0.6		12.6	1,000	390	700	310	S	16	n	limestone		365	clay
FL	Manatee Southwest	1982	1	reclaimed	irrigation	<5	1	1	1.0	1	10	520	690	170	S	16		carbonate	60,000		limestone, clay
FL	Marathon	1990	1	drinking water	PWS	11	0.25	0.25	1.0	0.25	100	368	427	59	S	16		limestone		700	dolomite
FL	Marco Lakes	1998	4	groundwater	PWS	>10	2	2	1.0	2	1,200	745	790	45	S	16	n	limestone	15,715	5500	clay

FL	Miami Beach	1998	1	groundwater	PWS	<5					0	10	33	12	S	30	n	sandstone		<150	none
FL	Miami–Dade Water & Sewer Dept. NW	1998	3	raw groundwater	raw water	3	5	5	1.0	714	2,800	550	1,250	700	S	30	n	limestone			
FL	Miami–Dade Water & Sewer Dept. SW	1999	0	raw groundwater	raw water																
FL	Naples	2009	0	groundwater	PWS	<5	1	2	0.5	2		1,080	1,350	270	ST	24	n	limestone	35,000 -200,000	2860	dolomite
FL	North Port	2004	1	reclaimed	irrigation	<3	1	1	1.0	1.5	120	583	650		PVC	12	n	limestone			
FL	NW Hillsborough	2003	1	reclaimed	irrigation	3	2	2	1.0	4	1,350	305	415	110	S	16	n	limestone	2,3000	787	dolomite
FL	Oldsmar	2012	n/a	reclaimed	irrigation							350	400	50				limestone			
FL	Orange County Utilities	2012	1	reclaimed	irrigation	3						1,100	1,200	100	FG	24	n	limestone		50	
FL	Palm Bay – formerly Port Malabar	1989	1	groundwater	PWS	>10	1	1	1.0	1	90	296	320	24	PVC	12	n	limestone	17,200	1320	dolomite
FL	Palmetto	2007	1	reclaimed	irrigation	0	1.2	1.2	1.0	1.2	200		520	520				limestone			limestone, clay
FL	PBC 3	1998	1	raw groundwater	raw	<6	1	1	1.0	1		1,055	1,200	145	S	18	y	sandstone		37000	dolomite
FL	PBC System 9 Hillsboro	2003	1	surface water	raw water	>2	5	5	1.0	5	900	1,000	1,225	225	S	14	n	limestone		4400	clay limestone, dolomite
FL	Peace River, Manasota Regional Water Supply Authority	1985	21	surface water	raw water	>20	1	1	1.0	18	1,600	500, 700, 900	110	72			n	limestone	5,000	700	dolomite
FL	Pinellas County (converted Class I wells)	1970	8	reclaimed	irrigation	<10	15	15	1.0	40		640	1,510		S	16 & 30	n	limestone			
FL	Polk	2102	1	reclaimed	irrigation	0						2,000	2,800	800	S			limestone			
FL	Punta Gorda – Shell Creek	1996, 2002	4	drinking water	PWS	10	1.4	1.4	1.0	5.6		764	933	169	S	12.25	n	limestone			
FL	Sanford	2008	1	PWS	PWS	>4	1	1	1.0	1.5		530	630	100	S	12		limestone		280	
FL	Sarasota Central	2001	0	reclaimed	irrigation		1	1	1.0	1			432	432							

(continued)

State	Project, Location	Estimated Start Year	Number of Wells in Project	Source of Water	Use of Water	Storage Cycles	Injection Cap (*mgd*)	Withdrawal Capacity (*mgd*)	Ratio In, Out	Peak Flow on Site (*mgd*)	Amount of Water Stored (*mg*)	Depth of Casing (*ft*)	Depth of well (*ft*)	Injection Horizon (*ft*)	Casing Material	Casing Diameter (*in.*)	Tubing and Packer?	Formation type	Transmissivity (*gpd, ft*)	TDS (*mg, L*)	Confinement
FL	Sarasota Venice Gardens	2012	3	reclaimed	irrigation	<5	0.5	0.5	1.0	1	60	500	700	200			n	limestone			
FL	Seminole	2004	1	reclaimed	irrigation	5	1	1	1.0	1	15	940	1,070	130	PVC	10.75	n	limestone		465	dolomite
FL	SFWMD Jupiter	1974	1	raw groundwater	raw groundwater	4	5	5	1.0	5	500	990	1,280	290	S	12	n	limestone		1980	dolomite
FL	St. Lucie Co SFWMD	1982	1	surface water	surface water	1	0.5	1.2	0.4	1.2	3.4	600	775	175	PVC	12 & 6	y	limestone		950	limestone
FL	City of St. Petersburg	2004	2	reclaimed	irrigation		1	1	1.0	1	700	490	600	110				limestone			
FL	Sunrise Springtree	2002	1	drinking water	PWS	5	1	1	1.0	1	270	1,110	1,270	160	S	16	n	limestone		6000	clay
FL	Tampa Hillsborough River	1991	0	surface water	raw	<5	0.6	0.6	1.0	1.2		466	535	69			n	limestone	500,000	1,300-4,000	limestone
FL	Tampa Rome Ave	1999	8	surface water	PWS	7	1.2	1.25	1.0	10	1,000	298	370	72	S	16	n	limestone	20,000	232	limestone
FL	West Palm Beach	1997	1	surface water	raw	5	3	8	0.4	8	520	970	1,200	230	S	24	n	limestone	100-135,000	5000	clay
GA	Dalton	2007	5	groundwater	industry													granite	ND		granite
GA	Flint River	2014	0	surface water	raw	0												limestone			
IA	Ankeny, Des Moines	1992	1	surface water	PWS	5	1	1.5	0.7		150	2,100	2,523		S	6	n	sandstone	12,800	200	
IA	Grimes	2012	1	groundwater	PWS	<5	1.1				27	2,110	2,602				n	sandstone			
IA	Jordan R, Des Moines	1989	1	surface water	PWS	5	1.5	3	0.5	3		2,185	2,515		S	6	n	sandstone	15,000 – 30,000	750	
IA	North Liberty	2009	1	groundwater	PWS	<5	1.2	1.7	0.7	1.7	27	1,575	1,673				n	sandstone			
ID	Boise, Micron Technologies	1998	3	process, surface	cooling UF	>10	1.4	2.1	0.7	11	900	600	1,219	16	S	8	n	basalt	30,000-60,000	200	silt, basalt
ID	Boise, United Water	1991	5		PWS	>10	0.3	0.3	1.0	1.1	15	70	220	4,7/92	S	10	n	basalt			

IL	Highland Park	study	0																		
KS	Republican R	2008	0	surface water	raw	0	0.15	10	0.0			9	50	21	S	9	n	alluvial	42,000		
KS	Wichita , Equus Beds	1997	4	surface water	raw	>5	10	10	1.0	10	70	60	130	70	S	18	n	alluvial			
MN	Joint Powers Water Board of Albertville, Hanover, and St. Michael	2007	1	surface water	raw	>5	0.3			0.45	5		504				n	alluvial			sedimentary
NC	Greenville	1996	1	surface water	PWS	<5	0.4	0.6	0.7	6	75	315 & 440	375 & 460	70	S	12		limestone			limestone
NC	Wilmington	2010	1	surface water	PWS	1	1.3	1.9	0.7	3.8	26 (135)	1,150	1,700	550	SS	24	n	alluvial	10,000		clay
NJ	Aqua New Jersey, Woolwich	2012	1	surface water	PWS	<2	0.2	0.6	0.3	1.6	10			0				sand			
NJ	Atlantic City MUA	2004	1	surface water	PWS	>10	0.5	1.7	0.3	1.7	400	722	780	58	S, SS	10	n	sand, clay			
NJ	Brick Township Municipal Utilities Authority	1996	1	surface water	PWS	>10	0.5	3	0.2	3	2,100	1,612	1,827	215	S	24 × 14	n	sand, clay			
NJ	Evesham Twshp Municipal Utilities Authority	1996	2	groundwater	PWS	>10	0.5	0.5	1.0	1.2	1,300	536	673	137	S	20 × 14	n	sand, clay			
NJ	S. Lakewood	2010	1	surface water	PWS	<5	2	2	1.0	2		1,400	1,621	221	SS	18 × 12	n	sand, clay			
NJ	Gordons Corner	1971	2	groundwater	PWS	>20	0.4	0.8	0.5	1.6	1,200	646	761	115	S	18 × 12	n	sand, clay			
NJ	Matchaponix	1989	1	surface water	PWS	>10	0.4	1.5	0.3	3	2.1	325	395	70	S, SS	19 × 12	n	sand, clay			
NJ	Aberdeen	1989	1	surface water	PWS	>10	1	1	1.0	1		500	560	60	S	24	n	sand, clay			
NJ	Jackson Twp	1993	1	surface water	PWS	>10	0.9	0.9	1.0	0.9		530	633	103	S	18 × 12	n	sand, clay			
NJ	Merchantville	2003	1	surface water	PWS	>10	1.5	1.5	1.0	1.5		107	185	78	S, SS	24 × 16	n	sand, clay			
NJ	United Water Manalapan	2006	1	surface water	PWS	>5	3	3	1.0	3		528	636	108	S, SS	20 × 14	n	sand			

(*continued*)

State	Project, Location	Estimated Start Year	Number of Wells in Project	Source of Water	Use of Water	Storage Cycles	Injection Cap (*mgd*)	Withdrawal Capacity (*mgd*)	Ratio In, Out	Peak Flow on Site (*mgd*)	Amount of Water Stored (*mg*)	Depth of Casing (*ft*)	Depth of well (*ft*)	Injection Horizon (*ft*)	Casing Material	Casing Diameter (*in.*)	Tubing and Packer?	Formation type	Transmissivity (*gpd, ft*)	TDS (*mg, L*)	Confinement
NJ	Monroe Township Municipal Utility Authority	1997	2	groundwater	PWS	>10	2	2	1.0	3	2,000	700 & 864	828 & 964		S, SS & PVC	14	n	sand			
NJ	Moorestown Department of Public Works	1998	1	surface water	PWS	>10	0.6	1	0.6	1	200						n	sand, clay			
NJ	Mt. Laurel Municipal Utility Authority	2003	1	surface water	PWS	>5	1.2	3	0.4	3	2,600		650	650		18 × 12	n	sand, clay			
NJ	NJAWC Smithville, Galloway	2009	1	surface water	raw	<5	0.5	0.9	0.6	0.9	40	509	552	43	S, SS	24 × 14	n	sand , clay			
NJ	NJAWC Western Division, Cherry Hill	1993	1	surface water	PWS	>10	1.1	1.1	1.0	1.7	2,000	264	353	89	S, SS	12	n	sand			
NJ	Swimming River	1988	1	surface water	PWS	>20	0.6	0.6	1.0	0	1,260	540	668	128	S	16	n	silt, sand	41,000–55,000	212	
NJ	United Water, Toms River	1999	2	groundwater	PWS	>10		1	0.0	3	2,100	1,535	1,756	221	S, SS	20 × 14	n	sand			
NJ	Wildwood	1968	4	groundwater	PWS	>30	0.4	0.4	1.0	1.3	164	212	328	116	S	12	n	sand	60,000		clay
NM	Alamogordo	1996	1	groundwater	PWS	1	0.8				21	750	850	100	S	18	n	limestone	1–10		
NV	Dayton Valley (Truckee Meadow)	2005	1	groundwater	PWS	<5	0.2	0.2	1.0	1	10,000	185	380	195	4 steel, 1 PVC	18, 4 inner tube	n	alluvial		420	
NV	Las Vegas Valley Water District	1987	87	varies, incl. reuse	raw	>20	1	6	0.2	>10	90,000	800	1,200	400	S	12, 16, 24	n	alluvial	83,000	467	
NV	N Las Vegas	1989	10	varies, incl. reuse	raw	>10	1	6	0.2	6.5	14,000	450	750	300	S	12, 14, 16, 20	n	alluvial	83,000		
NV	Truckee Meadows Water Authority Reno Recharge	1990	23	surface water	PWS	<5	0.4	0.4	1.0	0.8		110	260–820	700	S	18	n	alluvial		350	Bed rock

NV	Washoe County, Truckee	2008	4	surface water	PWS	<5	0.2	0.2	1.0	0.6	80		250	250	S	8	n	alluvial			Bed rock
NY	Bay Park	1968	1	PWS	PWS	19	0.3	0.6	0.5		>100	418	480	62	S	16	n	alluvial	1,000		
NY	East Meadow Reclaimed Water	2000	5	reclaimed	PWS	<5	0.5	0.5	1.0	2		65	96	31		12	n	sand	2,100		
OK	Oklahoma City, Blaine Recharge Demo Project	1972	5	surface water	raw	>20	1	1.5	0.7		46	750	985	260–430	S		n	alluvial			
OR	Baker City	2004	0	surface water	PWS	>5	1.1	1.5	0.7	2	745	740	800	60	S		n		1,000		
OR	Beaverton, Tualatin Valley Water District	1999	4	surface water	PWS	>10	0.8	0.8	1.0	3	428	204	450	246	S	12	n	basalt	65,500		basalt, clay
OR	Clackamas River Water	2000	1	surface water	PWS	<5	0.5	0.5	1.0	3	552	300	560	260	S	16	n	basalt	1,750		basalt, clay
OR	City of Dalles	2006	1	surface water	PWS	>10	0.28	0.28	1.0	0.28	273	500	2,001	1,501	S	16	n	basalt	1,300		basalt, clay
OR	Gannon Nursery	2013	0	surface water	irrigation	0												basalt			basalt, clay
OR	Hermiston	1994	0	surface water	raw																
OR	Madison Farms	2003	1	surface water	irrigation	>10	1	1	1.0	4	400		600	7.5	S		n		10,700		basalt, clay
OR	McCarty Ranch	2003	1	surface water	irrigation	>10	1	1	1.0	4	464		600	7.5	S		n		10,700		basalt, clay
OR	City of Pendleton	2003	2	surface via riverbank filtration	raw	<5	1.7	1.7	1.0	3	<400	630	700 & 714	70	S		n	basalt	35,300		basalt, clay
OR	Portland Water Bureau (Shore)	2003	7	surface water	PWS	<5	1.1–4	4		23.7	3,700	345 & 440	490 & 610	85	S		n	alluvial, sandstone	1,080		basalt, clay
OR	Salem, City of	1997	15	surface water	raw	>10	1.3	1.5	0.9	24.5	457	125	329	204				basalt	14,500		basalt, clay
OR	Sunrise Water, Happy Valley, Damascus	2007	1	surface water	PWS	>10	0.9	1.3	0.7	1.3	600	450	515	37–61	S		n	alluvial	1,400		basalt
OR	Tigard, City of	2001	2	surface water	raw	<10	1.4	1.4	1.0	12.5	605		1,225	12	S, SS	16	n	alluvial	9,800		basalt
OR	Tualatin, City of	2004	1	PWS	PWS	<10	1	2	0.5	17.2	2,700	580	920	340			n	alluvial	3,540		basalt
OR	Weyerhaeuser–Sycan Shop	2012	7	surface water	industrial																

(continued)

State	Project, Location	Estimated Start Year	Number of Wells in Project	Source of Water	Use of Water	Storage Cycles	Injection Cap (*mgd*)	Withdrawal Capacity (*mgd*)	Ratio In, Out	Peak Flow on Site (*mgd*)	Amount of Water Stored (*mg*)	Depth of Casing (*ft*)	Depth of well (*ft*)	Injection Horizon (*ft*)	Casing Material	Casing Diameter (*in.*)	Tubing and Packer?	Formation type	Transmissivity (*gpd, ft*)	TDS (*mg, L*)	Confinement
SC	Beaufort–Jasper Water & Sewer Authority	1995	3	surface water	PWS	>20	1.5	3	0.5	5	500	450	700	250	S	16	n	limestone			Dolomite
SC	Charleston	1998	1	surface water	raw	4	0.001	0.002			52	115 & 300	136 & 440	140	SS	6	n	limestone			
SC	Edistro Beach	2013	1	surface water	PWS								3,832	3,832			n	limestone			
SC	Georgetown	2010	3	surface water	PWS									0	S			limestone			
SC	Hilton Head	1988	1	surface water	PWS	<5	1	2	0.5	2.4	80			0		15		limestone	3070		
SC	Kiawah Island	1993	2	surface water	PWS	>5	0.6	0.6	1.0	0.6	60	250	300	50	S		n	sandstone			
SC	Mt Pleasant Waterworks & Sewer Commission	1992	3	surface water	PWS	<4	1	1	1.0	0.3	70		700	700	S		n	limestone			
SC	Myrtle Beach	1987	1	surface water	PWS	10	0.5	0.5	1.0	0	0	400 & 700	500 & 900	150	S		n	limestone			
SC	Myrtle Beach, Grand Strand WSA	1990	17	surface water	PWS	>20	0.5	1	0.5	8	500	700	900	200	SS	14	n	limestone	70		limestone
SC	Orangeburg	2012	2	surface water	PWS	3	2	3	0.7	6	6	325 & 640	480 & 870	160	SS	14	n	sands			
SC	South Island PSD	2012	1	reclaimed	irrigation	>4	1			1.2		191	618	427	S	12		limestone		280	
SD	Huron	1991	6	surface water	PWS	<5	0.1	0.2	0.5	0.2	10	60	120	60	S			alluvial	13,000	1400	
TN	Memphis Light Gas and Water	1995	1	ground	PWS	5	0.8	1.2	0.7	1.2		349	394	45	S, SS	10	n	sand		140	
TX	El Paso	1985	11	reclaimed	irrigation, raw	>10	2				400	632	861	229	S		n	alluvial			
TX	Guadelupe Blanco	2013	0																		
TX	Kerrville	1990	2	surface water	PWS	>10	1	1.5	0.7	2.5	12,000	485	571	86	S		n	sand			shale
TX	Laredo	1996	4	surface water	PWS	<5	0.2	0.2	1.0	0.1	0.325	260	420	160	S	6	n	sandstone	7.5	2000	silt, sand, clay

TX	New Brackfuls	2014	0																		
TX	San Antonio Water System	2004	33	groundwater	raw	>10	0.7			4	400	617	850	233	S		n	sand			
UT	Brigham City	1998	3	groundwater	PWS	>10	1	1	1.0	3	450	140 & 391	620	320	S	12, 20	n	alluvial	75,000		
UT	Leamington Town Culinary Water Well	2008	1	groundwater	PWS	<5	1.3	1.3	1.0	0.1	3,675	200	658	458	S	20	n	limestone	2 million	250	clay
UT	Jordan Valley Water District	1994	18	surface water	PWS	<5	0.7	0.7	1.0	12.8	170	400	1,000	600	S	20	n	alluvial	75,000-300,000		
UT	Salt Lake City, High Plains	2003	40	surface water	raw	<5	0.6	0.6	1.0	1	10	15	40	25		40	n	alluvial	107,000-232,000		
VA	Chesapeake	1987	1	surface water	PWS	>20	0.1	0.5	0.2	3	1,000	560	745	594–620 & 645–720	S	24	n	sand	9,500	212	
VA	Norfolk	1971	1	surface water	raw	3	0.4	1.5	0.3	10	20	896	976	80	FG, SS	8	n	sand	5,300	3000	
WA	Boise Wakulla Industrial	1991	4	surface water	raw	>10		2.1	0.0			200	450	250	5	12	n	basalt			basalt
WA	Kennimack, Richland	2008	0	PWS	PWS													basalt	80,000		
WA	Lakehaven, Tacoma	1991	2	surface water	raw	>10	0.3	5.4	0.1	1.5	10,000	250	300	50	S		n	basalt			
WA	Redmond	2002	0	surface water	raw	>2	0.2	0.2	1.0	0.2	10		200		S	16	n	basalt	350		unconfined
WA	Sammamish Water	2005	3	surface water	raw	<5	0.6	0.6	1.0	0.6	550		949		S	20	n				
WA	Seattle Public Utilities Highline	1989	3	surface water	PWS	>5	1	4	0.3	5.1	1,000	285	335	50	SS	16	n	alluvial	150,000 – 350,000	<200	
WA	Walla Walla	1991	2	surface water	PWS	4	2.3	4	0.6		87,000	138 & 810	245 & 1,220		S	524	n	basalt	5,500		basalt, clay
WA	White Salmon	2012	0	surface water	raw																
WA	Yakima	2002	1	PWS	PWS	2	1.7	3	0.6	90	120	876	1,163	287				basalt			
WI	Green Bay Utility	1998	1	surface water	PWS	<5	1	1.7	0.6	1.7	60	265	815	550	S	16	n	sandstone, limestone	14,000	250	silt
WI	Oak Creek	1998	2	surface water	PWS	<5	1	1.5	0.7	1.5	65	584	1,770	1,186	S	19	n	sandstone	12,000	485	silt

(continued)

State	Project, Location	Estimated Start Year	Number of Wells in Project	Source of Water	Use of Water	Storage Cycles	Injection Cap (*mgd*)	Withdrawal Capacity (*mgd*)	Ratio In, Out	Peak Flow on Site (*mgd*)	Amount of Water Stored (*mg*)	Depth of Casing (*ft*)	Depth of well (*ft*)	Injection Horizon (*ft*)	Casing Material	Casing Diameter (*in.*)	Tubing and Packer?	Formation type	Transmissivity (*gpd, ft*)	TDS (*mg, L*)	Confinement
WY	Cheyenne	2010	1	surface water	PWS	<5	0.2	0.2	1.0	0.2	32	315	375	60	S	10	n	sandstone			
WY	City of Gillette Lodgepole Creek	2001	0	surface water	irrigation																
	TOTAL SITES	204	726																		

[1] steel

[2] polyvinyl chloride

[3] stainless steel

[4] reverse osmosis

[5] public water system

[6] fiberglass

All data provided from best sources available at the time of collection or information provided. Numerous wells have little or no data available, or no response.

Data may have changed or been altered since collection.

Appendix B

Canadian ASR Experience

Canada has one ASR system, the Mannheim system, located in the city of Kitchener, Ontario, outside Waterloo. The system pumps raw Grand River water through a pipeline to the 16-mgd Mannheim Water Treatment Plant, where it is treated to potable standards through filtration and ultraviolet treatment and injected into a semi-confined, alluvial aquifer at the plant. Storage is designed to meet summer peak demands.

The well was constructed of a 16-in.-diameter stainless-steel well screen and a 17.4-in.-diameter PVC well casing. A silica sand pack was installed around the well screen and the remainder of the borehole opening was sealed with cement grout. A 100-hp submersible pump was installed below the well screen and connected to the surface with an epoxy-coated steel pipe.

The well was tested to simulate operational cycles (injection, storage, and recovery) using potable water from the water treatment plant starting in 1991. Three cycle tests were completed with progressively increasing storage volumes of 5, 25, and 50 million gallons. Storage times for each cycle were 3, 15, and 42 days, respectively. The ASR well was capable of injecting, storing, and recovering potable water from the aquifer at a rate of up to 63 L/sec. A fully operational system was permitted in 2005.

Now, according to the Region of Waterloo Water Supply and Distribution website, waters "from the Mannheim Artificial Recharge Facilities and seven other groundwater wells are blended with the treated surface water and then stored at the Mannheim Reservoir and Pumping Station which has a total storage capacity of 106 ML. Prior to the treated water being pumped into the distribution system, anhydrous ammonia is injected to convert free chlorine into monochloramines, a more stable disinfectant, that will last throughout the ditribution network."

This page intentionally blank.

Glossary

anisotropic aquifer An aquifer in which the aquifer formation does not transmit water equally in all directions (i.e., the horizontal and vertical permeability are not equal, causing the water to move preferentially in one direction with respect to the other two). Anisotropy is a property of a water-bearing formation in which hydrologic properties are non-uniform (vertically and or horizontally)

annulus The space between any piping, tubing, or casing and the piping, tubing, or casing immediately surrounding it

aperture The slot size of a slotted well screen

apparent specific capacity The ratio of the yield of each well to its drawdown (gpm/ft), used to plot the operational trend of each well

aquiclude A low-permeability geologic unit that forms either the upper or lower boundary of a groundwater flow system. Clay is a typical aquiclude material. This term is not often used in the industry. The term *confining unit* is more commonly used to describe aquiclude formations.

aquifer A geologic formation, or part of a group of formations, or a combination of formations, that is saturated and sufficiently permeable to transmit water to wells and springs, in sufficient quantities to be recoverable

aquifer recharge (artificial recharge) When water of a given quality is introduced into an aquifer at a point that is intended to allow water supplies to flow into a wellfield production zone

aquifer reclamation A program where large quantities of higher quality waters are injected into a compromised aquifer

aquifer storage and recovery (ASR) The management of water supplies for both potable (drinking water) and nonpotable uses, including meeting the needs of long-term growth demands, emergency conditions, or because of poor water quality

aquifer storativity The volume of water added to a unit horizontal area of the aquifer, based on rise in the water table elevation (i.e., the ability of the aquifer to increase its capacity to store water)

aquitard A low-permeability geologic unit that can store groundwater and also transmit it slowly from one aquifer to another. A leaky aquifer formation may be an aquitard. This term is not often used in the industry any longer. The term *semi-confining unit* is more common.

artesian aquifer An aquifer confined between less permeable materials from which water will rise above the bottom of the overlying confining bed if able to do so (see *confined aquifer*)

artesian wells Wells drilled into confined aquifers that are under pressure. In artesian wells, the piezometric level is above the ground surface, meaning that water could flow to the surface unaided.

artificial aquifers Aquifers created with technology involving the introduction of large quantities of treatable water into an aquifer or zone of an aquifer for retrieval down-gradient of the point of injection

attenuation The production of an output of lower magnitude than input, or a decrease in levels over time

biological clogging (fouling) The clogging of a well screen or other media as a result of microbiological growth

borehole (well bore) The hole caused by drilling

brackish water Water containing from 1,000 to 10,000 ppm of total dissolved solids

capacity The rate of flow delivered by a pump, in units such as gallons per minute, cubic feet per second, or barrels per hour. To calculate the power needed or the size of prime mover required to produce a desired capacity, the rate of flow and total dynamic head must be determined.

capillary fringe The subzone between the unsaturated and saturated zones. The capillary fringe occurs when a film of water clings to the surface of rock particles and rises in small-diameter pores against the pull of gravity.

cavitation The formation and collapse of water vapor bubbles in the flowing water

chemical clogging The clogging of a well caused by chemical reactions in the formation

coefficient of permeability A measure of the ease with which fluid is transported through a porous matrix (see *hydraulic conductivity*)

confined aquifer When an aquifer is overlain by a confining unit, the water in the aquifer is said to be confined. Typically clay or another impervious layer separates confined aquifers from the surface. Most confined aquifers are under some amount of pressure, but not always.

confining unit A low-permeability geologic unit that forms either the upper or lower boundary of a groundwater flow system. Clay is a typical confining unit material.

connate water Water entrapped in the interstices of sedimentary rocks at the time of deposition

development (of wells) The period of time in which a well is pumped continuously to rid it of the sand, silts, and other particles so the well is clean and the formation is no longer releasing material

dynamic head The resistance to flow produced by a system. Dynamic head is equal to the sum of static head, velocity head, and friction head.

effective pore space An indication of how much of the void space within the rock or soil is capable of transmitting water. This is important because some rock formations may have considerable pore (or void) space, but because the pores are not interconnected, the rock or soil may have difficulty transmitting water.

evaporation The transformation of water from the liquid phase to a vapor phase in the atmosphere

evapotranspiration The combination of evaporation and transpiration. Water less than 4 ft below the ground surface may be subject to evapotranspiration.

flowing artesian wells Artesian wells where the water level stands above the land surface

geophysical logging A technique to lower probes into the formation to retrieve electrical impulses that can be converted to useful aquifer data

groundwater Water contained in interconnected pores located either below the water table in an unconfined aquifer or in a confined aquifer.

hydraulic conductivity (commonly denoted as *K*) A measure of the ease with which fluid is transported through a porous matrix (also called *coefficient of permeability*). Hydraulic conductivity is a hydraulic property given in terms of a unit cube of material; it should not be confused with transmissivity, which is another hydraulic property, but is given in terms of a unit prism of an aquifer (see *transmissivity*).

hydraulic gradient A measure of the vertical change of the aquifer over a given distance from a stationary base (i.e., the slope of the aquifer formation)

hydrologic cycle The movement of water in the environment wherein water falls to the surface as rainfall, runs off over land to water bodies and/or infiltrates into the groundwater, and returns to the atmosphere via evaporation/transpiration, then returns as rainfall

injection horizon (injection zone) A geological formation, group of formations, or part of a formation receiving fluids through a well

intrusion The movement of saltwater into freshwater aquifers

isochlor An imaginary line on the earth's surface, as represented on a map, connecting all points of equal concentration of chlorides. It may be used to represent such points in groundwater, tidal estuaries, and other such bodies of water.

isotropic aquifer An aquifer in which the formation transmits water equally in all directions within the aquifer formation (generally considered to be a theoretical, ideal situation, rarely encountered in nature)

karstic formation A geological formation shaped by the dissolution of a layer or layers of soluble carbonate rock like limestone, dolomite, or gypsum

leaching The dissolution of solids and chemicals into water flowing through a porous sample

mechanical clogging The result of air entrapment or other physical factors that prevent movement of water to a well

net positive suction head (NPSH) The amount of pressure that prevents water from vaporizing, which can cause cavitation (the formation and collapse of water vapor bubbles in the flowing water) and damage a pump

packer test Method of collecting water quality samples allowing definition of the vertical distribution of water quality and hydraulic conductivity

piezometric surface The level to which water will rise in an aquifer under the influence of surface conditions. In a water table aquifer, the piezometric surface is the water table level. In a confined aquifer, the piezometric section may be significantly above the top of the rock formation if the aquifer is under pressure.

potential evapotranspiration The amount of evapotranspiration that would occur, assuming the soil moisture was adequate at all times. During drought conditions evaporation cannot occur, and so actual evapotranspiration may be less. Because of the disparity among amounts of rainfall throughout the year, atmospheric temperatures, and the freeze/thaw cycle, the actual evapotranspiration in the winter months is typically less that in summer months.

potentiometric surface The imaginary line where a given reservoir of fluid will equalize when allowed to flow

recharge areas Areas of porous surface soil that have a downward flow of percolating precipitation

redevelopment (of wells) Pumping, cleaning, or other measures to restore capacity

saltwater intrusion The movement of saltwater into a body of freshwater, either surface water or groundwater

semi-confining unit A low-permeability geologic unit that can store groundwater and also transmit it slowly from one aquifer to another. A leaky aquifer formation may be a semi-confining unit.

specific capacity The yield of a well, usually expressed in gallons per minute per foot of drawdown

specific discharge potential The amount of water that an aquifer could discharge given its hydraulic conductivity and piezometric head

specific injectivity The ratio of the injection rate accepted by the well to its water-level rise (gpm/ft), used to monitor the performance of an ASR well

specific yield The hydraulic property describing the available storage of a unit cube of material. It should not be confused with the storage coefficient, which is another hydraulic property describing available storage but is given in terms of a unit prism of aquifer (see *storage coefficient*).

spring The result of an aquifer being filled to the point that the water discharges over the land surface

static discharge head The distance measured vertically from the pump centerline to the water level in storage

static head The static suction head plus the static discharge head. To calculate static head, all measurements in pumping are vertical and the maximum drawdown is used as a reference. Measurements above this level are positive; those below, negative. The same measuring procedure can be used for both submersible and surface-mounted pumps.

static suction head The vertical measurement, in feet, of the distance from the water level in a well to the pump centerline

statigraphy A branch of geology that studies rock layers (strata) and layering (stratification), especially the distribution, deposition, and age of sedimentary rocks

stope leaching An alternative to or modification of conventional mining systems; it involves applying leach solutions to mineralized rock that has been fragmented by blasting ore in place, or by backfilling. The stope is the underground mining area.

storage coefficient The hydraulic property describing the available storage of a unit prism of aquifer. It should not be confused with the specific yield, which is another hydraulic property describing available storage but is given in terms of a unit cube of material (see *specific yield*).

submergence (commonly denoted as *S*) The measure from water level to introduction of air in a well

subsidence Downward movement of the land surface that can have a significant consequence on groundwater development from unconsolidated or highly friable rock formations

transmissivity (commonly denoted as *T*) The aquifer characteristic that is defined by the rate of flow per unit width through the entire thickness of an aquifer per unit gradient. This is valid only in two-dimensional flow. Transmissivity is a hydraulic property given in terms of a unit prism of an aquifer. It should not be confused with hydraulic conductivity, which is another hydraulic property but is given in terms of a unit cube of material (see *hydraulic conductivity*). Transmissivity can be found by multiplying the hydraulic conductivity (*K*) by the saturated aquifer thickness.

transpiration The process by which plants absorb water from the soil and release it to the atmosphere through their leaves

unconfined aquifer (water table aquifer) An aquifer where the water table is exposed to the atmosphere through openings in the overlying materials

underreaming The process of enlarging a section of wellbore beneath a restriction. The most frequently encountered restrictions are the internal diameter (ID) of the casing and the size of the wellhead. Both limit the maximum outside diameter (OD) of the tools that can pass through.

upconing The process by which saline water underlying freshwater in an aquifer rises upward into the freshwater zone as a result of pumping water from the freshwater zone

vadose zone The unsaturated portion of the aquifer

velocity head The height through which a buoy must fall freely to attain its velocity. In most cases, the velocity head is small and can be ignored.

Venturi diffuser A device that uses a parabolic converging section to centralize the flow from the nozzle into the Venturi throat

viscosity Combined with the fluid density, viscosity forms kinematic viscosity, which is important in determining the hydraulic conductivity of the soil or rock matrix.

void space The pore space within the soil or rock matrix that does not contain solid material

vug A small unfilled cavity in a rock

vugular formation A rock formation that contains significant connected vugs, cavities, or holes

water horsepower The work required to lift a weight of water to a defined height per unit of time (usually a second or a minute)

withdrawal test (slug test) Withdrawing a known volume of water into or out of a well and immediately measuring the rate at which the water level falls or rises back to static conditions

zone of discharge A predefined three-dimensional area in the ground around an installation where primary and secondary groundwater quality standards do not apply and effluents have an opportunity to diffuse or degrade somewhat before leaving the boundaries of the zone of discharge

Bibliography

Anderson, Michael K. 2012. ASR experience in Iowa. *Proceedings of the 57th Midwest Ground Water Conference.* October 1–2, 2012, Brooklyn Center, MN. Minnesota Ground Water Association. www.mgwa.org/meetings/2012_fall/aquifer_characterization/anderson.pdf. Accessed 3/5/13.

Arthur, J. D. 2004. Water-rock interactions during ASR and effects on water quality Aquifer Storage and Recovery. *IVAGWT/HC/FGS.* April 15-16, 2004, Tampa, Florida

AWWA. 2010. *Sustainable Water Resources: A Compendium of Issues and Trends CD-ROM,* F. Bloetscher, editor. Denver, CO: AWWA.

Barnes, Ferland and Associates. 2011. Expanded Executive Summary Orange County Utilities Aquifer Storage Recovery (ASR) System. Orlando, FL.

Bloetscher, F. and A. Muniz. 2010. Is ASR a viable strategy for emergency water supplies, *Opflow.* 36(8):24–27.

Bloetscher, F. and A. Muniz. 2010. Summary of ASR technology, *Sustainable Water Sources Conference Proceedings,* Albuquerque, NM, April 2010. Denver, CO: AWWA.

Bloetscher, F. 2004. Comparisons of results from modeling ASR system: Can modeling help with design of ASR systems? *GWPC Annual Meeting Proceedings.* Oklahoma City, OK: Groundwater Protection Council.

Bloetscher, Frederick and Albert Muniz. 2004. Can the results of modeling ASR systems answer long-term viability issues? *Environmental & Water Resources Institute (EWRI) Conference Proceedings,* Salt Lake City. UT. Reston, VA: ASCE.

Bloetscher, Frederick and Albert Muniz. 2004. Results of modeling brackish ASR systems. *2004 AWWA Water Sources Conference Proceedings,* Jan. 11–14, Austin, Texas. Denver, CO: AWWA.

Bloetscher, Frederick and Pamela Kenel. 2004. Comparisons of ASR practices in Florida and the Carolinas. *Florida Section AWWA Annual Conference Proceedings.* Hollywood, FL: FSAWWA.

Bloetscher, F., A. Muniz, and A.M. Gonzalez. 2002. Summary of AWWA Survey of ASR Practices in the United States, *Florida Section AWWA Annual Conference Proceedings.* Hollywood, FL: FSAWWA.

Bloetscher, Frederick and Gerhardt M. Witt. 1997. Aquifer storage and recovery as a means of resource management: A summary. *AWWA Annual Conference and Exposition Proceedings.* Atlanta, GA, June 1997. Denver, CO: AWWA.

Bloetscher, Frederick. 1996. Aquifer storage and recovery to meet peak water demands in southwest Florida. *CONSERV96 Conference Proceedings.* Denver, CO: AWWA.

Bowman, John. 2012. Aquifer storage and recovery an OASIS. Lakehaven Utility District, Washington. www.ecy.wa.gov/programs/wr/wstf/images/pdf/asr-bman.pdf. Accessed 3/5/13.

Brand, C. 2005. Colorado Springs' ASR Program. *Western Water Workshop Proceedings*, Gunnison, CO, July 28, 2005.

Brown, Donald L. and William D. Silvey. 1977. Artificial Recharge to a Freshwater-Sensitive Brackish-Water Sand Aquifer, Norfolk, Virginia Geological Survey Professional Paper 939, Washington, DC: United States Government Printing Office.

Brown, Phil and Justin Iverson. 2009. ASR optimization and expansion study City of Dallas, Oregon. OWRD Grant #GB0008-09. Lake Oswego, OR: Golder and Associates.

Bush, Peter W. 1988. *Simulation of saltwater movement in the Floridan aquifer system, Hilton Head Island, South Carolina.* U.S. Geological Survey water-supply paper 2331. Supt. of Docs, no.: I 19.13:2331. http://pubs.usgs.gov/wsp/2331/report.pdf. Accessed 3/5/13.

California Department of Water Resources (CDWR). 1983. Well completion report for 05S04W22M008S. Retrieved from files 4/20/13.

CDWR. 1986. Well completion report for 06SO4W21J003S. Retrieved from files 4/20/13.

CDWR. 1991. Well completion report for WCR485-48S. Retrieved from files 4/20/13.

CDWR. 1999. Well completion report for WCR715370. Retrieved from files 4/20/13.

CDWR. 2006. Local Groundwater Assistance Program, CA-program AB#03_Finalized_052006, www.water.ca.gov/lgagrant. Accessed 3/4/13.

Cardno/Entrix. 2011. Aquifer Recharge Using Reclaimed Water Collier County, FL, 2011 WaterReuse Seminar. info.ncsafewater.org/Shared%20Documents/Web%20Site%20Documents/Committees/WaterReuse/2011WaterReuseSeminar/2011ReuseSeminar_Waldron_ASRs.pdf. Accessed 3/5/13.

Carter, Janet M. 1996. Hydrologic data for 1994–96 for the Huron Project of the High Plains Ground-Water Demonstration Program. U.S. Geological Survey Open-File Report 96-555. Rapid City, SD: USGS.

Carter, Janet M. 1995. Hydrologic data through 1993 for the Huron Project of the High Plains Ground-Water Demonstration Program. U.S. Geological Survey Open-File Report 95-313 Rapid City, SD: USGS.

Castro, J.E. 1997. Aquifer storage and recovery: A water resources management alternative for Horry County, South Carolina. *Journal of Contemporary Water Research and Education*. 106(1).

CH2M Hill. 2004. Aquifer Storage and Recovery Demonstration Testing Operations Plan, Green Bay Water Utility. St. Louis, MO: CH2M Hill.

CH2M Hill. 2004. Aquifer Storage and Recovery Operations Plan, Oak Creek Water and Sewer Utility. St. Louis, MO: CH2M Hill.

CH2M Hill. 2007. Arsenic Mobilization in Two Suwannee Limestone Aquifer Storage Recovery Systems. Gainesville, FL: CH2M Hill.

CH2M Hill. 2002. Aquifer Storage and Recovery Hydrogeological Feasibility Study for the City of Pendleton, Oregon. Portland, OR: CH2M Hill.

CH2M Hill. 1999. Step 2 Report, Feasibility Investigation Aquifer Storage and Recovery. San Antonio, TX: CH2M Hill.

CH2M Hill. 1998. Engineering Report on the Construction and Testing of the Aquifer Storage and Recovery Facility at the City of Delray Beach's North Storage Reservoir. Deerfield, FL: CH2M Hill.

CH2M Hill. 1998. Engineering Report on the Construction and Testing of the Aquifer Storage and Recovery (ASR) System at the MDWASD West Wellfield. Deerfield, FL: CH2M Hill.

CH2M Hill. 1998. Engineering Report on the Construction and Testing of the Aquifer Storage and Recovery System at the North Storage Reservoir. Deerfield, FL: CH2M Hill.

CH2M Hill. 1998c. Construction and Testing of the Aquifer Storage and Recovery at the West Palm Beach Water Treatment Plant. Deerfield, FL: CH2M Hill.

CH2M Hill. 1998d. Aquifer Storage and Recovery, Preliminary Investigation and Feasibility Analysis, Step 1 Report. San Antonio, TX: CH2M Hill.

CH2MHill. 1997. Aquifer Storage and Recovery Hydrogeological Feasibility Study of Copper Mountain Basalt Aquifer, for Tualatin Valley Water District. Portland, OR: CH2M Hill.

CH2M Hill. 1997a. Engineering Report on the Construction and Testing of the Aquifer Storage and Recovery (ASR) System at the BCOES 2A Water Treatment Plant. Deerfield, FL: CH2M Hill.

CH2MHill. 1993. Aquifer Storage and Recovery, Feasibility Investigation, Phase IIB Full Scale Testing and Evaluation – Upper Guadalupe River Authority, Kerrville, TX. Austin, TX: CH2M Hill.

CH2M Hill. 1993. Engineering Report for the Boynton Beach Aquifer Storage and Recovery System. Deerfield, FL: CH2M Hill.

CH2M Hill. 1993. Engineering Report on the Construction and Testing of the Aquifer Storage and Recovery (ASR) System, Boynton Beach, FL. Deerfield, FL: CH2M Hill.

CH2M Hill. 1990. Interim Report for the Construction Florida Keys of the Aquifer Storage and Recovery Test Program. Deerfield, FL: CH2M Hill.

CH2M Hill. 1990. Interim Report on the Construction of the Florida Keys Aquifer Storage and Recovery Test Program Phase three. Deerfield, FL: CH2M Hill.

CH2M Hill. 1989. Aquifer Storage and Recovery, Feasibility Investigation, Phase IIA Monitoring Wells PZ1 – Upper Guadalupe River Authority, Kerrville, TX. Austin, TX: CH2M Hill.

CH2M Hill. 1989. Construction and Testing of the Aquifer Storage and Recovery (ASR) Demonstration Project for Lake Okeechobee, FL. Deerfield, FL: CH2M Hill.

City of Cocoa, FL. Aquifer Storage & Recovery Wells. www.cocoafl.org/index.aspx?NID=265. Accessed 3/5/13.

City of Greenville, NC. 2013. Greenville Utilities Aquifer Storage and Recovery. www.guc.com/About/AquiferStorageandRecovery.aspx. Accessed 3/5/13.

City of Salem, OR. 2012. Aquifer Storage. www.cityofsalem.net/Departments/Public-Works/Operations/WaterServices/Pages/AquiferStorage.aspx. Accessed 3/5/13.

City of Tigard, OR. 2012. Aquifer Storage and Recovery: Tigard Completes Construction of a 2nd Aquifer Storage and Recovery Well. www.tigard-or.gov/city_hall/departments/public_works/asr.asp. Accessed 3/29/13.

City of Tualatin, OR. 2013. Aquifer Storage and Recovery System. www.tualatinoregon.gov /publicworks/aquifer-storage-and-recovery-system. Accessed 3/5/13.

City of Wichita, KS. 2013. ASR Wichita's Aquifer Storage and Recovery Phase 2 Surface Water Treatment Plant – Equus Beds Aquifer Storage and Recovery Project www.wichitawaterproject.org/asr-phase-2.html. Accessed 3/5/13.

Colorado State Engineer. 2004. Colorado Division of Water Resources, Well Permit 259036. Denver, CO: Colorado Office of State Engineer.

Colorado State Engineer. 2006. Colorado Division of Water Resources, Well Permit 62850F. Denver, CO: Colorado Office of State Engineer.

Dick, Timothy A. 1993. Yakima Project, Bureau of Reclamation. www.usbr.gov/projects/ImageServer?imgName=Doc_1305647126617.pdf.

Dittus, R. 2008. Converting a supply well into an ASR well: Case histories. *Proceedings of the Annual Conference of the Pacific Northwest Section AWWA*.

Eaton, Larry. 2009. Aquifer Storage and Recovery (ASR): A Tool for Managing Oregon's Drinking Water. Kennewick, WA: GSI Water Solutions, Inc. www.oregonapwa.org/conference/fall09/Presentations/Aquifer.pdf. Accessed 3/15/13.

Eckhart, G. 2012. Aquifer Storage and Recovery. www.edwardsaquifer.net/asr.html. Accessed 3/5/13.

Emmert, Anthony A. 2010. Application for FCGMA Board Approval to Use City Well No. 29 as a Recharge Facility to Conduct an Emergency Injection/Storage Program, Oxnard CA. http://ventura.granicus.com/MetaViewer.php?view_id=45&clip_id=2085&meta_id=227920.Accessed 4/3/13.

Florida Department of Environmental Protection. 2013. Draft Permit to Richard Griswold, Destin Water Users, Inc. WACS ID 98505, from Joe Haberfeld.

Flournoy, Karen A. Undated. Draft Underground Injection Control Permit Number IAS153210004 Authorization to Operate a Class V Injection Well City of Grimes, IA. USEPA Region 7, Lenexa, KS. www.epa.gov/region07/public_notices/sdwa/pdf/grimes_asr_well_permit_draft.pdf. Accessed 3/15/13.

Foreman, Terry L. 2004. ASR Activities in the Western United States. *Aquifer Storage and Recovery IV Conference, Tampa, FL April 15, 2004.* www.dep.state.fl.us/geology/geologictopics/asr4/presentations/thursday_15/foreman_agwt4_15_04.pdf. Accessed 3/5/13.

Forrest, Glenn E. 2012. St. Johns River Water Management District, ASR Program Review, *FSAWWA Conference Proceedings*, Orlando, FL.

Foxworthy, B.L. 1970, Hydrologic Conditions and Artificial Recharge through a Well in the Salem Heights Area of Salem, Oregon. USGS Water Supply Paper 1594 F. Washington, DC: USGS

Gaisheng Liu, 2012. Report as of FY2011 for 2011KS113G: Aquifer Storage and Recovery in Near-Surface Aquifers: Development of a New Recharge Approach Using Small-Diameter, Low-Cost Wells, Kansas Geological Survey, Lawrence, KS: University of Kansas.

Geller, Doug. 2004. City of Portland Bureau of Water Works Columbia South Shore Well Field Aquifer Storage and Recovery Program – 2003–2004 Update. Portland, OR: Bureau of Water Works.

Germain, S. and Flynn, T. 2005. Aquifer Storage and Recovery Assessment, City of Kennewick, WRIA 31 Supplement to Water Storage Project, Prepared for WRIA 31 Planning Unit. Seattle, WA: Aspect Consulting.

Golder Associates. 2005. Aquifer Storage and Recovery Feasibility Study, City of Dallas, Oregon Water Treatment Plants Site. Lake Oswego, OR: Golder Associates.

Golder Associates. 2000. Technical Memorandum on Hydraulic Feasibility of Clacksamas River Water Well No. 1 for Aquifer Storage and Recovery. Redmond, WA: Golder Associates.

Golder Associates. City of Yakima ASR Pilot Test Preview Presentation to the Washington Department of Ecology ASR Rulemaking Technical Advisory Group. Redmond, WA: Golder Associates.

Golder Associates. 2008. Groundwater Storage in The Chimacum Creek Basin Draft Step 1 Report Screening of Prospective Sources and Recharge Sites March 17, 2008. Prepared for: Jefferson PUD #1 and the WRIA 17 Planning Unit P8399720cv01.indd 03/12/08. Redmond, WA: Golder Associates.

Golder Associates and Montgomery Watson. 1994. Hydrologic Feasibility Report – Groundwater Recharge for an Aquifer Storage and Recovery Pilot Program, City of Salem Department of Public Works. Lake Oswego, OR: Golder Associates.

Golder Associates and MWH. 2001. Phase I ASR Feasibility Report – City of Tigard, OR. Lake Oswego, OR: Golder Associates.

Grand Strand Water and Sewer Authority. 2012. Aquifer Storage and Recovery Wells. www.gswsa.com/products_and_services.cfm?page=19. Accessed 3/5/13.

Groundwater Geek. 2012. Walla Walla. www.groundwatergeek.com/asr-by-state/washington/city-of-walla-walla. Accessed 4/30/12.

Groundwater Management Associates. 2013. *What's New Aquifer Storage Recovery (ASR).* www.gma-nc.com/whatsnew.asp#Aquifer. Accessed 5/1/13

Groundwater Solutions, Inc. (GSI). 2003. Baker City Aquifer Storage and Recovery Feasibility Report. Portland, OR: GSI.

GSI. 2006. Deep Alluvial Aquifer – Aquifer Storage and Recovery (ASR) Feasibility Study. Prepared for Sunrise Water Authority. Portland, OR: GSI.

GSI. 2013. Madison Farms and McCarty Ranch ASR Projects Echo, Oregon. http://gsiwatersolutions.com/. Accessed 3/6/13.

GSI Water Solutions, Inc. and HDR, Inc. 2010. ASR-5 Well Report Boise White Paper, LLC. Wallula Mill Thermal ASR Project. Kennewick, WA: GSI Water Solutions, Inc.

Hazen and Sawyer. 2010. City of Naples Well Completion Report for Aquifer Storage and Recovery Well No. 1. Boca Raton, FL: Hazen and Sawyer.

HDR Engineering. 2006. Comprehensive Water System Plan Update-Final Report. Prepared for the City of Walla Walla, WA.

Hemenway, Courtney and Paul Grundemannz. 2002. ASR as a Water Management Tool for Highlands Ranch, Colorado. *Proceedings for Ground Water/Surface Water Interactions, AWRA Summer Specialty Conference,* July 1–3, 2002. http://infohouse.p2ric.org/ref/33/32646.pdf. Accessed 3/5/13.

Hemenway, C. 2004. Letter to Valois Shea, RE: Request for Rule Authorization for Consolidated Mututal Water Company.

Hilton Head Public Service District. 2012. Hilton Head (SC) Public Service District Groundwater Crisis Management: Past – Present – Future. Hilton Head, SC.

Hilton Head Public Service District. 2012. Hilton Head Water Supply & Saltwater Intrusion Presentation to Beaufort County Council, September 2012. Hilton Head, SC.

Hurlow, Hugh, M. Lowe, M. Matyjasik, and Paul Gettings. 2011. The Weber River Basin Aquifer Storage and Recovery Pilot Project Special Study 136. Salt Lake City, UT: Utah Geological Survey, A Division Of Utah Department Of Natural Resources.

Idaho Department of Water Resources. 2010. Water Permit Report. Water Right 63-31183 (Micron Technology). Boise, ID.

Idaho Department of Water Resources. 2008. Injection Well Permit 52-W-156-001. Boise, ID: Micron Technology.

Izbicki, John A., C.E. Petersen, K.J. Glotzbach, L.F. Metzger, A.H. Christensen, G.A. Smith, D. O'Leary, M.S. Fram, T. Joseph, and H. Shannon. 2010. Aquifer storage recovery (ASR) Of chlorinated municipal drinking water in a confined aquifer. *Applied Geochemistry*, 25:1131–1152.

Jackson, Tom. 2012. City of Punta Gorda Utilities Department Inter-Office Memorandum to Howard Kunik, City Manager. March 12, 2012. Subject: Aquifer Storage and Recovery (ASR). Punta Gorda, FL.

Kansas Dept. of Health and Environment. 2006. Underground Injection Control Area Permit KS-05-079-001 to City of Wichita (Equus Beds ASR). Topeka, KS: Dept. of Health and Environment.

Kenel, Pamela P., R. Moresi, and F. Bloetscher. 2004. Survey of ASR systems in the Carolinas. *SCEP Conference Proceedings*. Myrtle Beach, SC: SCEP.

Kimley Horn. 1996. Aquifer Storage and Recovery Demonstration Project for the City of Miami Beach, Phase II Final Report. Miami, FL: Kimley Horn & Associates.

Lin, L. Yu. 2001. Memphis and Shelby Counties, Tennessee Case Study #9. *Regional Studies in Environmental Science*. New York: McGraw Hill Educational. www.mhhe.com/biosci/pae/environmentalscience/casestudies/case9.mhtml. Accessed 5/3/13.

Livingston, E.C. and S.T. Finch. 1997. Aquifer Storage and Recovery Study for the City of Alamogordo, New Mexico, Water Issues of Eastern New Mexico, Get Your Kicks on Route 66. New Mexico Water Resources Research Institute, October, 1997. wrri.nmsu.edu/publish/watcon/proc42/Livingston_Finch.pdf. Accessed 2/14/13.

Lowe, Mike, H.A. Hurlow, and M. Matyjasik. 2003. The Weber River Basin Aquifer Storage and Recovery Project Open-File Report 419 September 2003. Salt Lake City, UT: Utah Geological Survey, A Division of Utah Department of Natural Resources.

Lower Republican River Stakeholder Advisory Committee. 2010. Recommendations for improvements to the water supply, management, efficiency and conservation in the Lower Republican River Basin. Topeka, KS: Kansas Water Office.

Malcolm Pirnie, Inc., ASR Systems, LLC, and Jackson, Sjoberg, McCarthy & Wilson, LLP. 2011. An Assessment of Aquifer Storage and Recovery in Texas. Texas Water Development Board Report # 0904830940. Austin, TX: TWDB.

Maliva, R. J., R.F. Griswold, and M.M. Autrey. 2013. Prototype for a Reclaimed Water Aquifer Storage and Recovery System: Benefits and Operational Experiences, *FWRJ*, 65(2):54.

Memphis Light, Gas and Water. 1995. Application for Authorization for Class V Underground Injection Well, submitted to the Tennessee Department of Environment and Conservation Division of Water Supply. Memphis, TN: Memphis Light, Gas and Water.

Mills, Suzanne, K., C. Peterson, and J. Damerel. 2009. Aquifer Storage & Recovery in California, Current Projects and Issues, Groundwater Resources Associate of California, 27th Biennial Groundwater Conferences, October 7, 2009. https://sunsite.berkeley.edu/WRCA/WRC/pdfs/GW27thMills.pdf. Accessed 3/5/13.

Minnesota Department of Health. 2009. Joint Powers Water Board brings aquifer storage technology to Minnesota. *Waterline.* www.health.state.mn.us/divs/eh/water/com/waterline/featurestories/jpwb.html. Accessed 3/4/13.

Missimer, Thomas M., Fredrick Bloetscher, and Charles W. Walker. 2002. Use of Aquifer Storage and Recovery Technology to Improve Membrane Treatment Plant Efficiency - Collier County, Florida. *Desalination,* 87(1–3): 269–280.

Missimer, Thomas M. and R. Maliva. 2010. Aquifer Storage and Recovery and Managed Aquifer Recharge Using Wells: Planning, Hydrogeology, Design, and Operation. Houston, TX: Schlumberger.

Mittelstet, Aaron R., M.D. Smolen, G.A. Fox, and D.C. Adams. 2011. Comparison of aquifer sustainability under groundwater administrations in Oklahoma and Texas. *Journal of the American Water Resources Association,* Vol. 47, No. 2.

Monterey Peninsula Water Management District. 2006. *Santa Margarita Test Injection Well Aquifer Storage and Recovery System, Operation and Maintenance Manual.* Monterey, CA: Monterey Peninsula Water Management District.

Monterey Peninsula Water Management District. 2006. Seaside Groundwater Basin Aquifer Storage and Recovery (ASR), Phase 1 request for funds. Monterey, CA: Monterey Peninsula Water Management District. www.mpwmd.dst.ca.us/Mbay_IRWM/ASR/Proj_Sum_1.pdf. Accessed 3/2/13.

Monterey Peninsula Water Management District. 2010. Monte Vista Water District Aquifer Storage & Recovery Program, www.mvwd.org/ps.yourwater.cfm?ID=167. Accessed 3/2/13.

Morris, Sharon, and G. Carpenter. 2009. Aquifer Recharge Experiences and Technologies: Thriving in the Desert, *Arizona-Baja California Sur Partnership for Water Sustainability Workshop.* La Paz, Baja California, Mexico, April 24, 2009. http://www.azwater.gov/AzDWR/Hydrology/Library/documents/AZ-BCS_RechargePresentation.pdf. Accessed 3/5/13.

Muniz, Albert, A.M. Gonzalez, and F. Bloetscher. 2003. Summary of AWWA survey of ASR practices in the United States. *Florida Water Resource Conference Proceedings,* Tampa, Florida. St. Cloud, FL: FWRC.

Murray, Smith and Associates. 2003. Aquifer Storage and Recovery Limited License Application and Pilot Test Work Plan. Portland, OR: Murray, Smith and Associates.

MWH. 2002. Springtree (WTP) Aquifer Storage and Recovery (ASR) System Cycle Testing Report (Cycles 1 through 6, UIC Permit No. 127248-001-UC. Sunrise, FL: MWH.

MWH. 2002. City of Fort Lauderdale, Five Ash Water Treatment Plant Aquifer Storage and Recovery System Well Completion Report. Sunrise, FL: MWH.

Nevada Division of Environmental Protection. 2008. Dayton ASR Renewal - UIC Permit Renewal UNEV2006200. Lyon County Utilities Department, Aquifer Storage and Recovery. Reno, NV: Nevada Division of Environmental Protection.

Nevada Division of Environmental Protection. 2008. Golden Valley ASR Renewal - UIC Permit Renewal UNEV91008. Washoe County Dept. Water Resources / Utilities Division, Aquifer Storage and Recovery. Reno, NV: Nevada Division of Environmental Protection.

Nevada Division of Environmental Protection. 2008. Truckee Meadows ASR Renewal - UIC Permit Renewal UNEV92200. Truckee Meadows Water Authority, Aquifer Storage and Recovery. Reno, NV: Nevada Division of Environmental Protection.

Norton, S.B. 2007. Quantifying the near borehole geochemical response during aquifer storage and recovery: Application of "push-pull" analytical techniques to ASR cycle testing. Thesis. Gainesville, FL: University of Florida.

Orange County Water District. 2013. Groundwater Replenishment System, Fountain Valley, CA. www.gwrsystem.com. Accessed 3/2/13.

Oregon Department of Environmental Quality (DEQ). 2005. Water Pollution Control Facilities Permit for Class V Stormwater Underground Injection Control System, City of Portland Permit No. 102830. Salem, OR: DEQ.

Oregon Water Resources Department. 2013. Aquifer Storage and Recovery (ASR) Limited License #014, for ASR testing to: Madison Farms. Salem, OR: DEQ.

Oregon Water Resources Commission. 2011. Aquifer Storage and Recovery (ASR) Limited License #003, for ASR Testing to: Clackamas River Water; Water Resources Department. Salem, OR: DEQ.

Oregon Water Resources Commission. 2009. Aquifer Storage and Recovery (ASR) Limited License #010, for ASR Testing to: City of Tualatin; Water Resources Department. Salem, OR: DEQ.

Oregon Water Resources Commission. 2008. Aquifer Storage and Recovery (ASR) Limited License #016, for ASR Testing to: City of Pendleton; Water Resources Department. Salem, OR: DEQ.

Oregon Water Resources Commission. 2007. Aquifer Storage and Recovery (ASR) Limited License #013, for ASR Testing to: McCarty Ranch. Salem, OR: DEQ.

Oregon Water Resources Commission. 2007a. Aquifer Storage and Recovery (ASR) Limited License #012, for ASR Testing to: Sunrise Water Authority, Water Resources Department. Salem, OR: DEQ.

Oregon Water Resources Commission. 2006. Aquifer Storage and Recovery (ASR) Limited License #011, for ASR Testing to: City of Dallas Water Resources Department. Salem, OR: DEQ.

Oregon Water Resources Department. 2006. Aquifer Storage and Recovery (ASR) Limited License #005 The Oregon Water Resources Commission issues this limited license for ASR Testing to City of Tigard. Salem, OR: DEQ.

Oregon Water Resources Commission. 2001. Aquifer Storage and Recovery (ASR) Limited License #004, For ASR Testing to: City of Portland; Water Resources Department. Salem, OR: DEQ.

Oregon Water Supply and Conservation Initiative (OWSCI). 2009. Underground Storage Inventory: Clackamas River Water ASR Oregon Underground Storage: Clackamas River Water ASR. Salem, OR: Oregon Water Resources Department.

OWSCI. 2009. Underground Storage Inventory: Madison Farms ASR and AR, Oregon Underground Storage: Madison Farms ASR and AR. Salem, OR: Oregon Water Resources Department.

OWSCI. 2009. Underground Storage Inventory: Oregon Underground Storage: Tigard, ASR. Salem, OR: Oregon Water Resources Department.

OWSCI. 2009. Underground Storage Inventory: Oregon Underground Storage: Madison Farms ASR and AR. Salem, OR: Oregon Water Resources Department.

OWSCI. 2009. Underground Storage Inventory: Oregon Underground Storage: Sunrise Water Authority ASR. Salem, OR: Oregon Water Resources Department.

OWSCI. 2009. Underground Storage Inventory: Oregon Underground Storage: City of Pendleton ASR. Salem, OR: Oregon Water Resources Department.

OWSCI. 2009. Oregon Water Supply and Conservation Initiative Inventory of Potential Below Ground Storage Sites. www.oregon.gov/OWRD/LAW/docs/owsci/OWSCI_GW_study_text.pdf. Accessed 3/15/13.

Osborn, N.J. and E. Eckenstein. 1997. Blaine Gypsum Groundwater Recharge Demonstration Project Data Report. Oklahoma Water Resources Board Technical Report 97-4. Oklahoma City, OK.

Palm Beach County Water Utilities Department. 2003. Eastern Hillsboro Canal WTP #0 Aquifer Storage and Recovery Well. West Palm Beach, FL: Palm Beach County Water Utilities Department.

Palm Beach County Water Utilities Department. 2003. Water Treatment Plant #3 Multipurpose Floridan Aquifer Storage and Recovery Well, Well Construction and Operational Testing Request. West Palm Beach, FL: Palm Beach County Water Utilities Department.

Peterson, C. 2007. Aquifer Storage & Recovery in California, Benefits and Considerations, *26th Biennial Groundwater Conferences,* Sept. 19, 2007. Sacramento, CA: Groundwater Resource Association of California.

Petersen, Christian E. and Kenneth Glotzbach. 2005. Aquifer Storage and Recovery for the City of Roseville: A Conjunctive Use Pilot Project. WEFTEC 2005. Alexandria, VA: WEF.

Petkewich, M.D., D.L. Parkhurst, K.J. Conlon, B.G. Campbell, and J.E. Mirecki. 2004. *Hydrologic and geochemical evaluation of aquifer storage recovery in the Santee Limestone/ Black Mingo aquifer, Charleston, South Carolina, 1998–2002: U.S. Geological Survey Scientific Investigations Report 2004–5046*, http://pubs.water.usgs.gov/sir2004-5046. Accessed 5/24/04.

Pool, R. and Eychaner. 1999. Measurements of aquifer storage change and specific yield using gravity surveys. *Groundwater*, 33(2):425.

Poole, B., L. Wiseman, and B.L. Turner. 2011. Seminole County Markham WTP Storage and Recovery Project Summary. August 31, 2011.

Pueblo Water Resources. 2012. Technical Report for Notice of Intent SWRCB Water Quality Order 2012-0010 City of Tracy Aquifer Storage and Recovery Project Pueblo Water Resources, Ventura, CA. www.swrcb.ca.gov/rwqcb5/board_decisions/adopted_orders/general_orders/2012-0010-dwq_noas/2012-0010-dwq-rb5s-0002.pdf. Accessed 3/5/13.

Pyne, David. 2007. Overview of Aquifer Storage Recovery in the USA and the World. Presentation to Conference on Aquifer Storage Recovery and Artificial Recharge in Oregon, November 7, 2007. http://water.oregonstate.edu/asr/presentations/Pyne.pdf. Accessed 3/15/13.

Pyne, R.D. 2005. *Groundwater Recharge and Wells: A Guide to Aquifer Storage Recovery*, 2nd Ed. Gainesville, FL: CRC Press.

Rask, B.M. 2005. Letter to V. Shea, RE: Request for Rule Authorization for a Class V injection well. Littleton, CO: JC Halepaska and Associates.

Reese, R. 2002. Inventory and Review of Aquifer Storage and Recovery in Southern Florida. US Geological Survey Water-Resources Investigations Report 02-4036. Tallahassee, FL: USGS.

Roseville, Calif. 2010. City of Roseville Aquifer Storage & Recovery. http://acwi.gov/swrr/Rpt_Pubs/wef_session107/107_xalt.pdf. Accessed 2/15/13.

Rydman, D. 2012. Lessons learned from and aquifer storage and recovery program. *JAWWA*, 104(9):52.

San Antonio Water System. 2012. Twin Oaks - Aquifer Storage & Recovery. www.saws.org/your_water/waterresources/projects/asr.cfm. Accessed 3/5/13.

San Antonio Water System. 2012a. Aquifer Storage & Recovery. http://www.saws.org/your_water/waterresources/2011_update/asr.cfm. Accessed 3/5/13.

Sheng, Zhuping. 2005. An aquifer storage and recovery system with reclaimed wastewater to preserve native groundwater resources in El Paso. *Texas Journal of Environmental Management*. 75(2005):367–377.

Skomrucha, S. 2009. Letter to K. Nickerson, RE: United Water Delaware Inc. Filing the Certification of Adequate Water Supply. Wilmington, DE: United Water Delaware, Inc.

Slusser, T. 2010. Reissuance and Modification of Injection Well Permit W10700040 (Greenville Utilities Commission). Raleigh, NC: NC Division of Water Quality.

Slusser, T. 2013. Modification of Cycle Testing Plan - Injection Well Permit W108000149 (Cape Fear Public Utilities Authority). Raleigh, NC: NC Division of Water Quality.

Sophocleous, Marios. 2009. Aquifer Storage and Recovery and the Lower Republican River Valley, Kansas, Kansas Geological Survey Open-File Report 2009-18. Kansas Geological Survey. Lawrence, KN: University of Kansas.

Thomas, J.M., W.A. McKay, E. Cole, J.E. Landmeyer, and P.M. Bradley. 2000. The fate of haloacetic acids and trihalomethanes in an aquifer storage and recovery program, Las Vegas, Nevada. *Groundwater*, 38(4):605–614.

Thomas, Kevin, Robert Q. Oaks, Paul Inkenbrandt, Walid Sabbah, and Mike Lowe. 2011. Open-Cache Valley Principal Aquifer Storage and Recovery Site Assessment: Phase I File Report 579. Salt Lake City, UT: Utah Geological Survey, a Division of Utah Department of Natural Resources.

Todd Engineers. 2012. Gloria Way Water Well Production Alternatives Analysis and Water Security Feasibility Study City of East Palo Alto, California. Alameda, CA: Todd Engineers

United States Geological Survey (USGS). 1995. Highline Well Field Aquifer Storage and Recovery Project Seattle, Washington – Final Report. Boise, ID: US Department of the Interior Bureau of Reclamation Snake River Area Office – West and Pacific Northwest Regional Office.

USGS. 2002. Inventory and Review of Aquifer Storage and Recovery in Southern Florida. Water Resources Investigations Report 02-4036. Tallahassee, FL: USGS.

USGS. 2012. Equus Beds Groundwater Recharge Project. ks.water.usgs.gov/studies/equus/. Accessed 3/5/13.

Utah Water Quality Board. 2011. Utah Water Quality Board Class V Area Permit Underground Injection Control (UIC) Program Permit Number UTU-03-AP-173E18B Permit Modification and Renewal Issued to: Brigham City Corporation. Salt Lake City, UT: Utah Water Quality Board.

Utah Water Quality Board. 2008. Utah Water Quality Board Class V Area Permit Underground Injection Control (UIC) Program Permit Number UTU-27-5-E1AAR076 Permit Issued To: Leamington Town Corporation. Salt Lake City, UT: Utah Water Quality Board.

Utah Water Quality Board. 2010. Utah Water Quality Board Class V Area Permit Underground Injection Control (UIC) Program Permit Number UTU-35-AP-38871EB (Formerly UTU500005) Permit Modification and Renewal Issued to: Jordan Valley Water Conservancy District. Salt Lake City, UT: Utah Water Quality Board.

Valencic, C. vs. Department of Environmental Protection and Sarasota County. OGC Case No. 01-1284, DOAH Case No. 01-3535 Final Order.

Vorissis, Mary, Jim Lozier, Paul Franks, Ken Ortega, and Anthony Emmert. 2004. Oxnard California's Groundwater Recovery Enhancement and Treatment (Great) Program. http://ebookbrowsee.net/paleoresourceinventory-impactassessmenttechrpt-pdf-d28079107. Accessed 3/11/13.

Wallin, E.T. 2006. Cancellation of ASR Limited License, 104, by Licensee (Portland), Final Order. Salem, OR: Oregon Water Resources Board.

Washington Department of Ecology. 2006. Aquifer Storage and Recovery Permit RI 28083P, Lakehaven Utility. Bellevue, WA: Dept. of Ecology.

Washington Department of Ecology. 2005. Aquifer Storage and Recovery Permit RI 28192P, Sammamish Plateau Water and Sewer District. Bellevue, WA: Dept. of Ecology.

Washington Division or Water Management. 1991. Water Well Report 201607 Yakima Well. Bellevue, WA: Dept. of Ecology.

WE&T Magazine. 2007. Plant profile: Elmendorf Twin Oaks Aquifer Storage and Recovery Facility Location: Elmendorf. *WE&T,* 19(11). www.wef.org/publications/page_wet.aspx?page=plant&id=47231/2. Accessed 3/5/13.

Weber Basin Water Conservancy District. 2011. Weber River Basin Aquifer Storage and Recharge Project – Updates. www.weberbasin.com/index.php/environmental/mitigation-projects/128. Accessed 3/5/12.

Wiemiller, Patrick N. 2013. Report to the City Council – Approval of Agreement. City of Fresno Agenda Item, Fresno, CA.

This page intentionally blank.

Index

S

T

U

V

W

Z

AWWA Manuals

M1, *Principles of Water Rates, Fees, and Charges*, #30001
M2, *Instrumentation and Control*, #30002
M3, *Safety Management for Water Utilities*, #30003
M4, *Water Fluoridation Principles and Practices*, #30004
M5, *Water Utility Management*, #30005
M6, *Water Meters—Selection, Installation, Testing, and Maintenance*, #30006
M7, *Problem Organisms in Water: Identification and Treatment*, #30007
M9, *Concrete Pressure Pipe*, #30009
M11, *Steel Pipe—A Guide for Design and Installation*, #30011
M12, *Simplified Procedures for Water Examination*, #30012
M14, *Backflow Prevention and Cross-Connection Control: Recommended Practices*, #30014
M17, *Installation, Field Testing, and Maintenance of Fire Hydrants*, #30017
M19, *Emergency Planning for Water Utilities*, #30019
M20, *Water Chlorination/Chloramination Practices and Principles*, #30020
M21, *Groundwater*, #30021
M22, *Sizing Water Service Lines and Meters*, #30022
M23, *PVC Pipe—Design and Installation*, #30023
M24, *Planning for the Distribution of Reclaimed Water*, #30024
M25, *Flexible-Membrane Covers and Linings for Potable-Water Reservoirs*, #30025
M27, *External Corrosion Control for Infrastructure Sustainability*, #30027
M28, *Rehabilitation of Water Mains*, #30028
M29, *Water Utility Capital Financing*, #30029
M30, *Precoat Filtration*, #30030
M31, *Distribution System Requirements for Fire Protection*, #30031
M32, *Computer Modeling of Water Distribution Systems*, #30032
M33, *Flowmeters in Water Supply*, #30033
M36, *Water Audits and Loss Control Programs*, #30036
M37, *Operational Control of Coagulation and Filtration Processes*, #30037
M38, *Electrodialysis and Electrodialysis Reversal*, #30038
M41, *Ductile-Iron Pipe and Fittings*, #30041
M42, *Steel Water-Storage Tanks*, #30042
M44, *Distribution Valves: Selection, Installation, Field Testing, and Maintenance*, #30044
M45, *Fiberglass Pipe Design*, #30045
M46, *Reverse Osmosis and Nanofiltration*, #30046
M47, *Capital Project Delivery*, #30047
M48, *Waterborne Pathogens*, #30048
M49, *Butterfly Valves: Torque, Head Loss, and Cavitation Analysis*, #30049
M50, *Water Resources Planning*, #30050
M51, *Air-Release, Air/Vacuum, and Combination Air Valves*, #30051
M52, *Water Conservation Programs—A Planning Manual*, #30052
M53, *Microfiltration and Ultrafiltration Membranes for Drinking Water*, #30053
M54, *Developing Rates for Small Systems*, #30054
M55, *PE Pipe—Design and Installation*, #30055
M56, *Nitrification Prevention and Control in Drinking Water*, #30056
M57, *Algae: Source to Treatment*, #30057
M58, *Internal Corrosion Control in Water Distribution Systems*, #30058
M60, *Drought Preparedness and Response*, #30060
M61, *Desalination of Seawater*, #30061
M63, *Aquifer Storage and Recovery*, #30063
M65, *On-Site Generation of Hypochlorite*, #30065

www.ingramcontent.com/pod-product-compliance
Lightning Source LLC
LaVergne TN
LVHW082344200926
848481LV00012B/3446